CTBUH Awards
Best Tall Buildings

A Global Overview of 2014 Skyscrapers

Antony Wood, Steven Henry & Daniel Safarik

Council on Tall Buildings and Urban Habitat

Routledge
Taylor & Francis Group

LONDON AND NEW YORK

Book Design & Layout: Marty Carver

First published 2014 by Routledge

2 Park Square, Milton Park, Abingdon, Oxfordshire OX14 4RN
52 Vanderbilt Avenue, New York, NY 10017

Routledge is an imprint of the Taylor & Francis Group, an informa business

First issued in paperback 2019

British Library Cataloguing in Publication Data
A catalogue record for this book is available from the British Library

Library of Congress Cataloging in Publication Data
A catalog record has been requested for this book

ISBN: 978-1-138-84289-2 (hbk)
ISBN: 978-0-367-37819-6 (pbk)

Acknowledgments

The CTBUH would like to thank all the organizations who submitted their projects for consideration in the 2014 awards program.

We would also like to thank our 2014 Awards Jury for volunteering their time and efforts in deliberating this year's winners and finalists.

About the CTBUH

The Council on Tall Buildings and Urban Habitat is the world's leading resource for professionals focused on the inception, design, construction, and operation of tall buildings and future cities. A not-for-profit organization based at the Illinois Institute of Technology, Chicago, with an Asian office at Tongji University, Shanghai, the group facilitates the exchange of the latest knowledge available on tall buildings around the world through events, publications, research, working groups, web resources, and its extensive network of international representatives. Its free database on tall buildings, The Skyscraper Center, is updated daily with detailed information, images, data, and news. The CTBUH also developed the international standards for measuring tall building height and is recognized as the arbiter for bestowing such designations as "The World's Tallest Building."

Contents

Foreword	6
Introduction	8
CTBUH Best Tall Building Awards Criteria	19

Best Tall Building Americas
Winner:
Edith Green-Wendell Wyatt Federal Building, *Portland*	22

Finalists:
The Point, *Guayaquil*	28
United Nations Secretariat Building, *New York City*	32

Nominees:
4 World Trade Center, *New York City*	36
Magma Towers, *Monterrey*	38
MuseumHouse, *Toronto*	40
Peninsula Tower, *Mexico City*	42
Territoria El Bosque, *Santiago*	44
The Godfrey, *Chicago*	46
Torre Costanera, *Santiago*	48
500 Lake Shore Drive, *Chicago*	50
1812 North Moore, *Arlington*	50
Concord Cityplace Parade, *Toronto*	51
Courtyard & Residence Inn, *New York City*	51
Couture, *Toronto*	52
K2 at K Station, *Chicago*	52
NEMA, *San Francisco*	53
The John and Frances Angelos Law Center, *Baltimore*	53
The Peter Gilgan Centre, *Toronto*	54
Torres del Yacht, *Buenos Aires*	54
YooPanama Inspired by Starck, *Panama City*	55
ZenCity, *Buenos Aires*	55

Best Tall Building Asia & Australasia
Winner:
One Central Park, *Sydney*	58

Finalists:
8 Chifley, *Sydney*	64
Abeno Harukas, *Osaka*	68
Ardmore Residence, *Singapore*	72
FKI Tower, *Seoul*	76
IDEO Morph 38, *Bangkok*	80
Sheraton Huzhou Hot Spring Resort, *Huzhou*	84

The Jockey Club Innovation Tower, *Hong Kong*	88
Wangjing SOHO, *Beijing*	92

Nominees:
41X, *Melbourne*	96
171 Collins Street, *Melbourne*	98
Academic 3, *Hong Kong*	100
Albert Tower, *Melbourne*	102
Anhui New Broadcasting & TV Center, *Hefei*	104
Baku Flame Towers, *Baku*	106
Changzhou Modern Media Center, *Changzhou*	108
China Merchants Tower, *Shenzhen*	110
Fake Hills, *Beihai*	112
Guangzhou Circle, *Guangzhou*	114
Habitat, *Melbourne*	116
Jinao Tower, *Nanjing*	118
Kent Vale, *Singapore*	120
L'Avenue, *Shanghai*	122
OLIV, *Hong Kong*	124
Shanghai Arch, *Shanghai*	126
Xiamen Financial Centre, *Xiamen*	128
ASE Centre Chongqing R2, *Chongqing*	130
Asia Square, *Singapore*	130
China Resources Building, *Hong Kong*	131
DBS Bank Tower, *Jakarta*	131
Fortune Plaza Phase III, *Beijing*	132
Infinity, *Brisbane*	132
Jinling Hotel Asia Pacific Tower, *Nanjing*	133
One AIA Financial Center, *Foshan*	133
RMIT Swanston Academic Building, *Melbourne*	134
The Capital, *Mumbai*	134
The Gloucester, *Hong Kong*	135
The Pakubuwono Signature, *Jakarta*	135

Best Tall Building Europe
Winner:
De Rotterdam, *Rotterdam*	138

Finalist:
DC Tower, *Vienna*	144

Nominees:
6 Bevis Marks, *London*	148
10 Brock Street, *London*	150
AvB Tower, *The Hague*	152

CalypSO, *Rotterdam*	154
Exzenterhaus Bochum, *Bochum*	156
Fletcher Hotel Amsterdam, *Amsterdam*	158
Maslak Spine Tower, *Istanbul*	160
Solaria, *Milan*	162
The Tower, One St George Wharf, *London*	164
Tour Carpe Diem, *Paris*	166
E' Tower, *Eindhoven*	168
Grand Office, *Vilnius*	168
One Angel Square, *Manchester*	169
Solea, *Milan*	169

Best Tall Building Middle East & Africa

Winner:

Cayan Tower, *Dubai*	172

Nominees:

BSR 3, *Tel Aviv*	178
Champion Tower, *Tel Aviv*	180
Portside, *Cape Town*	182
The Landmark, *Abu Dhabi*	184
22 Rothschild Tower, *Tel Aviv*	186
Conrad Hotel, *Dubai*	186
Rosewood Abu Dhabi, *Abu Dhabi*	187
World Trade Center Doha, *Doha*	187

Urban Habitat Award

Winner:

The Interlace, *Singapore*	190

Finalist:

NEO Bankside, *London*	196

Nominee:

Gramercy Residences, SkyPark, *Makati*	200

10 Year Award

Winner:

Post Tower, *Bonn*	204

Finalists:

Taipei 101, *Taipei*	210
Torre Agbar, *Barcelona*	210
Uptown Munchen, *Munich*	211
Highlight Towers, *Munich*	211

Time Warner Center, *New York City*	212
Bloomberg Tower, *New York City*	212
Tower Palace Three, *Seoul*	213

Innovation Award

Winner:

BioSkin	216

Finalist:

Active Alignment	220

Nominees:

DfMA and Digital Engineering for Tall Buildings	222
LiftEye	224
Steel Fiber Reinforced Concrete	226

Performance Award

Winner:

International Commerce Center, *Hong Kong*	230

Finalist:

Jin Mao Tower, *Shanghai*	234

Nominee:

Darling Quarter, *Sydney*	236

Lifetime Achievement Awards

Lynn S. Beedle Award, *Douglas Durst*	240
Fazlur R. Khan Medal, *Peter Irwin*	246
CTBUH 2014 Fellows	252

Awards & CTBUH Information

CTBUH 2014 Awards Jury	253
Review of Last Year's CTBUH 2013 Awards	254
Overview of All Past Winners	260
CTBUH Height Criteria	264
100 Tallest Buildings in the World	267

Index

Index of Buildings	272
Index of Companies	273
Image Credits	277
CTBUH Organizational Structure & Members	279

Foreword

Jeanne Gang, *2014 Awards Jury Chair*

It has been my pleasure to serve once again as Awards Jury Chair for this year's CTBUH Best Tall Building Awards. During my tenure, I've been consistently impressed with the depth and range of the juries' conversations. It is incredibly reassuring to recognize that despite our differences of discipline, geography, or language, we find so many commonalities when united in the appreciation of good design. Sharing ideas and analysis with such highly accomplished professionals and colleagues has been truly energizing.

It is always difficult and perhaps inherently unfair to judge a building through the brief process of an awards program when the actual project represents years of hard work by its design team. This year we re-examined and adjusted the process undertaken by the jury to allow more time for reflection and deeper inquiry into the projects' performance and innovation. Prior to selecting the regional Finalists, jurors were given an extra week to review the submissions, as well as the technical evaluations provided by our newly appointed Technical Jury. The extended timeline also provided the opportunity to solicit and review any additional information the jury felt was vital to the selection process and to request further feedback from the Technical Jury. These modifications served to strengthen an already thorough process, ensuring not only exemplary finalists throughout all of the categories but also careful attention to how we on the jury define and measure excellence.

While admirable buildings were submitted from every region, projects from Asia were undeniably compelling. With so many quality buildings to choose from, the jury found it difficult to select just a few. Ultimately, we decided to include a larger-than-usual sampling of Finalists from Asia, in order to recognize some of the most successful tall buildings in recent history.

Overall this year, the projects most appreciated by the jury propelled the tall building typology into new arenas, looking beyond pure height to their impact on urban environments. The expanded spatial aspects and connections between indoors and outdoors as demonstrated by IDEO Morph 38, Bangkok, and the Ardmore Residences, Singapore, both Finalists in the Asia & Australasia category, were especially promising. Projects such as The Interlace, Singapore, recognized as both a Finalist in the Asia & Australasia category and as the Winner of the inaugural Urban Habitat Award, have literally inverted conventional thinking about tall buildings. The way in which The Interlace and other projects engage the ground and the environment in new ways is exemplary. And the jury was particularly impressed with vegetated, solar-responsive tall building projects like One Central Park, Sydney, the Asia & Australasia Winner, which plays a role in urban ecology.

Having witnessed firsthand the progress and progressiveness of this building type, even just over the past two years, I am ever-more excited about the future of tall buildings and the transformative potential of the ideas they incubate. This year's entries show that innovation is everywhere, yet there is also much to improve. Herein lies the challenge: if we really want to radically improve tall buildings, we need to continue to push ourselves and our clients to measure success through understanding the building's energy performance. Unless we as designers and jurors prioritize, disclose, and track this information, we are missing a major point of impact and our process remains flawed. Let us continue to discuss the idea of energy performance and together find ways to encourage forward movement on this front.

Thank you to my fellow jurors Sir Terry Farrell; David Gianotten; Saskia Sassen; David Scott, who served as Technical Jury Chair; Thomas Tsang Wai Ming; and of course Antony Wood, whose experience and passion is appreciated by all. Thanks also to this year's technical jurors Guo-Qiang Li, Nengjun Luo, Douglas Mass, Paul Sloman, and Peter Williams. And finally, my deep appreciation goes to Steven Henry at the CTBUH for once again tirelessly stewarding this program.

Introduction

In 2014, it seems that the tall building community has finally recovered from the shock of the 2008 financial crisis, which had delayed or canceled many projects and had a chilling effect on investment in the immediate years following. But 2014 was not a year in which the built evidence demonstrated the industry simply shaking off the dust and resuming just as before. Many of the tall building projects of the boom era that never came to be – overwrought, gaudy, disconnected from place – were perhaps best left on paper. In 2014, there

are now strong signs that a return to largesse would not be a sustainable strategy. There seems to be a dawning recognition that, in order to deliver transformative, let alone viable, projects, some first principles about the way we build have to be re-examined.

Only a tiny selection of the tall buildings completing in the past 12 months or so have received CTBUH awards this year. But the value of throwing the spotlight on these projects is that doing so exposes themes that are reflective of, or have wider implications for, the industry as a whole. For in these projects we can see the vestigial signs of an emerging consciousness: one that recognizes the value of designing with, instead of against, nature – and the responsibility to do so. We see not a blind faith in technology, but a shrewd fusion of simple, time-tested principles with digital validation, creating solutions that have the potential to carry us through difficult conditions that surely await us in the near future. We see data being "liberated" and put to use in ever-expanding ways, so that as we ascend higher, we assume less. We see forms as inventive as ever, but with a renewed sense of purpose. And we see the achievements of our best practitioners as a reminder of why we build in the first place.

One of the strongest trends evident in the projects completing this year was the number of high-quality refurbishments and environmental upgrades of existing tall buildings, to become more energy-efficient, more functional, and safer. Establishing a precedent

The Americas Winner, the Edith Green-Wendell Wyatt Federal Building, and Americas Finalist United Nations Secretariat, were particularly outstanding in their refurbishment, because they reflect the commitment of national and trans-national government bodies to demonstrate the power of reinvestment in existing buildings when it comes to the environment. The Edith Green-Wendell Wyatt project is a dramatic conversion of a concrete-frame, 1970s office building into a light-filled, strategically shaded contemporary facility that engages its surroundings. Meanwhile, the UN Secretariat project restored the external appearance of one of the Modern movement's greatest icons, while substantially improving the mechanical, safety and communications systems, light penetration, and arrangement of the internal spaces.

Private industry has often railed against government regulation of its activities in the name of curbing climate change and greenhouse gases, and it's no secret that some of the suppliers of raw materials and hardware for tall buildings have been first to the front lines in such protests. As such, it is critical that governments take the steps to show that they are willing to undertake the degree of change they seem to be demanding of private concerns. And this isn't just a case of "self-sacrifice" for the betterment of the environment. In retrofit cases such as Edith Green-Wendell Wyatt, once-dismal and outdated real estate has become more valuable, and begins paying for itself almost immediately, not just in lower energy

for renovating tall buildings is clearly important to the industry, as there is no template for demolishing buildings of the height we are now seeing constructed when they begin to underperform. Nor can we expect that there will be much tolerance for the environmental consequences of demolishing massive buildings and wasting their embodied carbon, or continuing to operate them with outmoded and inefficient technology.

But we must be honest and admit that these buildings do have a service life that ends at some point. Now the second major wave of tall buildings – which, with their thin envelopes and tight engineering, have proved to perhaps be more difficult to retrofit than the masonry and punched-opening buildings of the first generation – has reached that point. However, at least two projects in this book prove that second-generation skyscrapers can be retrofitted for another half-century's operation, at least. Still, we are running out of excuses for not considering how the buildings we construct today might be continuously optimized or feasibly altered tomorrow.

bills, but in the higher productivity of its occupants. For all the exposure that operating energy costs get in the building industry, for most companies it is staff/salaries that are the number-one cost. Thus, if they're happier, healthier, and more productive, that's a bottom-line argument that is hard to counter.

Just as importantly, there now exists the beginning of a template for dealing with aging tall building stock. Many commercial mid-century office buildings share characteristics with this year's modernization awardees – low floor-to-ceiling heights, poor insulation, poor natural light penetration, outdated cellular office plans, and embedded, poorly performing HVAC systems. Many property owners are now facing a crisis in how to convert their underperforming assets from the Modernist era into contemporary Class-A office space that meets today's high occupant and regulatory standards. Through projects such as Edith Green-Wendell Wyatt and the UN Secretariat, governments now have shown that the "talk" can be "walked." Now it's the corporates' turn to do similar.

And so it is that "green" building design and retrofits have entered the architectural vanguard, and have been working their way up in scale to the tall buildings we are accustomed to working on and in. An even more recent, and sometimes even more sensually stunning development, is the incorporation of vertical greenery in tall buildings. Early pioneers from the 1990s include Germany's Commerzbank and Chile's Consorcio, and many of the CTBUH Best Tall Buildings regional winners of the past few years incorporate some form of sky lobby with planted greenery, such as The Bow, Calgary, or planted, habitable roof areas, such as the Pinnacle @ Duxton, Singapore. But in recent years, green walls have been taken, quite literally, to new heights, particularly in Southeast Asia, where the tropical climate is exceptionally conducive to plant growth. In fact, the CTBUH has researched this trend extensively, and released a comprehensive technical guide on the subject, *Green Walls in High-Rise Buildings*, this year.

Bangkok's Ideo Morph 38 project, included both in the Green Walls guide and in this volume, for instance, actually uses a panelized system of mesh with creeping vines as an element of architectural enclosure, providing aesthetic, shading, and cooling benefits up to the 32nd floor of one of the towers.

At One Central Park, Sydney, the 2014 Best Tall Building Asia & Australasia, technological sophistication in the form of directional heliostats cantilevering from the tower's peak regulates the sunlight that reaches the ground as well as the greenery-festooned shelves that cover all faces of the tower. This is not mere window-dressing; the unprecedented level of green implementations on this project not only provide shading, driving reductions in energy consumption, they also facilitate an intangible but critical new asset in the form of a public park. This inverts the traditional interpretation of tall buildings as constructions that are additive to, rather than subtractive from, the public realm. To

Opposite: Asia & Australasia Finalist Ideo Morph 38, Bangkok; creeping vines grow up the side of the building to the 32nd floor

Top: Asia & Australasia Winner One Central Park, Sydney; extensive use of greenery on all façades of the building

Bottom: Asia & Australasia Finalist Abeno Harukas, Osaka; the tallest building in Japan provides public green spaces throughout the building

add needed new housing and new public space that consumes an exemplarily low amount of energy – that is an achievement worth celebrating and studying.

Asia & Australasia Finalist Abeno Harukas, Osaka, the tallest building in Japan, takes a different approach; less obvious perhaps, but just as valid. Its green assets are very much "inside" the building, yet they are also shared with the public. At the top levels is a public garden and peripheral walkway, open to fresh air but also protected from stiff winds by a glass façade, thus combining a fantastic view with a stellar green asset. Elsewhere in the building, intermediate setbacks are similarly outfitted with refreshing, open, and yet protected patches of greenery. This is in addition to the amazing concentration of such a diverse range of amenities as retail, museum, hotel, office, school, hospital, and park in one small urban footprint over a busy railway station. Abeno Harukas pushes the boundaries of what tall buildings can achieve on several different levels.

As design technology has advanced over the years, it has become increasingly common to see new interpretations of the skyscraper form. Not all of these can be justified; many proposed during the building boom in advance of the financial crisis of 2008 carried little merit other than being sculptural curiosities. In today's more chastened climate, however, we now see a resurgence of unusual shapes, the best of which are elegant, distinctive,

and disciplined expressions of the building's mission, and often as not, are directly site- and environment-related, in addition to being aesthetically interesting.

The Europe Winner, De Rotterdam, is a deceptively straightforward project that appears as a reinterpretation of a multi-tower project from the High Modern era in North America, but then reveals myriad subtleties as one's viewing angle changes, and as day changes to night. The Netherlands' now largest building breaks down its scale by shifting the upper portions of its three towers slightly off-center, which seems a curious and minor intervention until nighttime backlighting reveals the structural acrobatics required to accomplish this. Is it one building, or three, or six? Is it a sail or a lantern? If a basic box form, sliced and set askew can yield so many possibilities, we can be confident that, in the hands of talented designers, there is much as-yet unseen life left in the typology.

Asia & Australasia Finalist Sheraton Huzhou Hot Spring Resort is unmistakable on the horizon. By day it is a serene mirage of a horseshoe forming a figure-eight shape with its reflection in the water; by night a glowing, buzzing icon of possibility. The cultural importance of the lakeside location is emphasized by the annular shape of the building, and yet its form is not purely symbolic. The shape of the hotel also provides maximal natural light for each room, as well as commanding views.

The seemingly willful undulating texture of the Ardmore Residence in Singapore – somewhere between the grille of a 1930s Duisenberg auto, a prehistoric monolith, and a futuristic wind instrument – is also a carefully chosen response to its site. The curvilinear, sculpted form optimizes views and maximizes shading to minimize heat gain, while capitalizing on opportunities for prefabricated, energy-saving construction, even though its appearance is as a continuously carved single piece.

Perhaps the most site-responsive of the ardently sculptural entries this year is also one of the most visually stunning – the Middle East & Africa Winner, Cayan Tower, Dubai. This tour-de-force of structural engineering, twisting through 90 degrees along its 306-meter rise, is actually a strikingly simple concept, turning repeating floor plates through slight rotations up to dizzying height and effect. It thus achieves iconicity while providing quality views and protection for its occupants from wind-driven sandstorms, by way of an economical construction strategy that doesn't give a hint of compromise.

One of the most important aspects of the CTBUH awards this year is the introduction of a new awards category: the Urban Habitat Award. You'll recognize the phrase as the second half of our organization's name, and may have wondered, "What is 'urban habitat,' exactly, and what do tall buildings have to do with it?" It turns out others have asked this question in one form or another, and quite frankly, it is often in criticism of the industry's relentless pursuit of height and iconicity sometimes to the detriment of those who still live close to the ground.

The Urban Habitat Award has been introduced to close the gap in thinking between "tall building design" and "urban planning." The CTBUH believes that tall buildings cannot be judged solely on their height or functional performance as singular entities. Their contribution to the urban realm must go beyond their iconic effect on the skyline. They must ultimately be judged on how they

integrate with cities at a human level. This new award thus recognizes significant contributions to the urban realm, in connection with tall buildings. In particular, it highlights buildings that demonstrate a positive contribution to the surrounding urban environment, add to the social sustainability of both their immediate and wider settings, and represent design influenced by context, both environmentally and culturally.

The quality and inventiveness of the responses to the inaugural Urban Habitat call for submittals was high. The inaugural winner, The Interlace, Singapore, accomplishes much of what the "Urban Habitat" part of the CTBUH mission seeks to encourage – the creation of quality urban space and communal space at height, the creative interpretation of the skyscraper typology, and the intensive integration of activities throughout the vertical and horizontal planes. Here again, first looks are deceiving – what seems a "crazy quilt" of tall buildings laid on their sides is actually a tightly ordered rendition of the vertical city concept appropriate to its site. In addition, every horizontal external surface is maximized for communal recreation space, creating perhaps the best realized version of the 1970s Japanese metabolist ideas yet seen. Moreover, it is suggestive of just how much freedom we actually have in organizing our cities of the very near future.

The adjective "high-performance" is so frequently appended to all manner of things from footwear to deodorant, that it is easy to become desensitized to its use with respect to buildings. What does it mean when

it comes to these giant objects in which we spend nearly our entire waking lives and which contribute so substantially to our footprint on the earth?

This year, the CTBUH has established the Performance Award for tall buildings, because all of the attendant effects of building production and operation – including the tendency of awards programs to recognize "design intent" before it has been proven – are amplified in tall buildings.

The Performance Award presents a conundrum, because performance can be assessed along many fields, including environment, energy use, occupant satisfaction, and economics. Ideally all of these are positively intertwined. But the conundrum persists, largely because of the industry's reticence about sharing building performance data. The new CTBUH Performance Award seeks to change this deficiency in our industry.

The winner and finalist of this year's inaugural award, the International Commerce Centre (ICC), Hong Kong and the Jin Mao Building, Shanghai, epitomize the multifaceted nature of "performance" in tall buildings. At ICC, the building's 98 percent occupancy level – including some of the world's most prominent financial institutions – is a testament to its excellent management as well as design. Here, sharing and transparency are watchwords. The building's intensive tenant-management relationship hinges on making energy-use data and reporting a key form of currency,

resulting in top-tier energy performance. Again, beneath all of the sophisticated measuring techniques is a simple proposition – mutual trust and information sharing leads to mutual benefit, as energy savings are channeled into more high-touch service and premium rents.

The Pudong district of Shanghai was mostly a low-rise warren of warehouses when the Jin Mao Building opened in 1999, offering high-end office, hotel and convention space. About to become the shortest of the "three sisters," a trio that includes the Shanghai World Financial Center and Shanghai Tower, the Jin Mao nevertheless stands tall in terms of its computerized energy and air-quality management systems. Technology is once again reinforced by the human factor – each month building managers hold an energy-consumption analysis meeting to ensure the 15-year-old tower is still on a positive course.

Now in its second year, the CTBUH 10 Year Award recognizes buildings that have delivered value and

performance over the past decade. This is a purposely broad category, and it encompasses aspects of many of the annual awards. The 10 Year Award can recognize buildings that have contributed to the urban realm, dealt with social issues, stood out in terms of occupant satisfaction, continued to outpace the market with technical or engineering performance, or satisfied many other criteria, ideally many at once.

The winner of this year's 10 Year Award, the Deutsche Post Tower in Bonn, Germany, is recognized for setting

precedents that many other buildings today have a hard time equaling: major energy savings are derived from natural ventilation and decentralized heating and cooling, and its pioneering use of a double-façade system provides fresh air while preventing drafts. It is also one of the best examples of a corporate tall building living up to its owners' desire that the building embody its values.

If anything can be said about the entries, and in particular the winner and finalist, of the CTBUH Innovation Award, it is that solutions need not be complex in order to be effective. For all the industry talks about the value of BIM, computational fluid dynamics and parametric modeling, it is refreshing to see that we still have within us an intuitive sense of what will work. Having said that, the solutions profiled in this book were doubtless validated with all manner of sophisticated computational devices, as well they should be.

The winner of the Innovation Award, BioSkin, as implemented in the NBF Osaki building in Tokyo, revives an ancient Japanese practice for cooling the environment – spreading water on hot surfaces and dusty streets – by incorporating it within a the façade shading strategy, for a hugely necessary antidote to the contemporary problem of the urban heat island effect. Using the simple principles of evapo-transpiration, BioSkin simultaneously relieves cooling loads for the building to which it is attached, and reduces urban temperatures in the surrounding area, while diverting site rainwater from overtaxed storm sewers.

To counteract issues of non-uniform deformation of asymmetric tall buildings, Active Alignment – a Finalist in the Innovation category – is a hydraulically assisted, calibrated update on the time-worn practice of fitting shims until the table stops wobbling. This could have tremendous implications for future projects, normalizing not only the settlement that inevitably occurs during construction, but also in operations, such as compensating for seismic and wind loading in a method significantly less obtrusive than thousand-ton mass tuned dampers taking up space in the proverbial "attic."

Both innovations are reminders of how much hard work actually goes into solutions that seem obvious only in retrospect.

If it seems that there is a recurring theme across all of these award winning projects and innovations, it is that in this technologically driven age, it must never be forgotten that buildings are made by, and for, people. That is why some of the most intuitive-seeming solutions still take years of hard work to arrive at. It's why placing greenery at heights where the environment has historically been treated as something to "keep out" has now become a best practice. And it's why, each year, we recognize two examples of human achievement that remind us of what we can achieve, and how much hard work is still to be done.

The Lynn S Beedle award this year goes to Douglas Durst of the Durst Organization, who, as a building developer, has shown faith and devotion to the city of New York as few others have, through the oil and debt crises of the 1970s, through the Wall Street booms and busts, and beyond the terrible legacy of the 2001 terrorist attacks. Something about the need for people to congregate in a great city, and the need for a great city to have equally great buildings, has inspired Durst to push forward with energy- and faith-saving investments alike, as in projects such as 4 Times Square (1999) to the Bank of America Tower (2009), and the soon-to-be-complete One World Trade Center. The return of New York to the forefront of the skyscraper vanguard currently has much to do with individuals like Douglas Durst.

A patience and fortitude of a different kind characterizes the Fazlur Khan Award 2014 winner, Peter Irwin, who has intensively studied and tested the performance of countless tall buildings in wind tunnels and through computer modeling. His firm, RWDI, is now one of the world's most respected authorities on wind engineering, and the Irwin Sensor, which tests

the comfort of pedestrians at the foot of tall buildings (there is the human factor again) is used at labs around the world. In Irwin lies a belief that every project is a valuable learning tool, even if it is never built. It is in that spirit of amassing knowledge that CTBUH moves forward with all of its awards, publications and research initiatives.

What, then, have we learned from this year's awards submissions, and what do they tell us about the future? If these tall buildings, and the people behind them, are in fact representative, then we have reason to be optimistic about the future. We can envision, with confidence, a future in which the spaces "in between" – at the ground, at the midpoint, at junctions and on the roof – are as important as the "big gesture" in the skyline. We can foresee a future in which tall buildings exist not just as static structures or engineering marvels, but as living, changing contracts between owner and occupant, and between the building and city scales. We can imagine a future in which the most practical design feature in a tall building is also its most beautiful, and the most cost-effective design is also the most awe-inspiring. And we can see a day in which tall buildings give new vitality to their urban surroundings, gathering up the "ordered chaos" of city life and commuting its vibrancy, sounds, smells and humanity to the sky, making something entirely new in the process. Thank you for supporting the Council on Tall Buildings and Urban Habitat in this endeavor.

Opposite Top: 10 Year Award Winner Post Tower, Bonn; an early precedent for environmentally conscious tall building design

Opposite Bottom: Innovation Award Winner BioSkin; rainwater runs through the pipes which cover the façade of the NBF Osaki Building, Tokyo. The evapo-transpiration of the rainwater naturally cools the micro-climate surrounding the building

CTBUH Best Tall Building

Awards Criteria

The Council on Tall Buildings and Urban Habitat initiated its Awards Program in 2001, with the creation of the Lynn S. Beedle Lifetime Achievement Award. It began recognizing the team achievement in tall building projects by issuing Best Tall Building Awards in 2007, to give recognition to projects that have made extraordinary contributions to the advancement of tall buildings and the urban environment, and that achieve sustainability at the highest and broadest level. It issues four regional awards each year, and from these "regional" awards, one project is awarded the honor of overall "Best Tall Building Worldwide," which is announced at the annual awards ceremony.

The winning projects must exhibit processes and/or innovations that have added to the profession of design and enhanced cities and the lives of their inhabitants. Some of the criteria for submission are outlined below. It is important to note that, with the exception of the first point (regarding completion date eligibility), a project does not necessarily need to meet every listed criteria. Submissions should demonstrate strengths in areas that are applicable:

1) The project must be completed (topped out architecturally, fully clad, and at least occupiable) no earlier than the 1st of January of the previous year, and no later than the 1st of October of the current awards year (e.g., January 1, 2013 and October 1, 2014 for the 2014 awards).

2) The project advances seamless integration of architectural form, structure, building systems, sustainable design strategies, and life safety for its occupants.

3) The project exhibits sustainable qualities at a broad level:
Environment: Minimize effects on the natural environment through proper site utilization, innovative uses of materials, energy reduction, use of alternative energy sources, and reduced emissions and water consumption.
People: Must have a positive effect on the inhabitants and the quality of human life.
Community: Must demonstrate relevance to the contemporary and future needs of the community in which it is located.
Economic: The building should add economic vitality to its occupants, owner, and community.

4) The project must achieve a high standard of excellence and quality in its realization.

5) The site planning and response to its immediate context must ensure rich and meaningful urban environments.

6) The contributions of the project should be generally consistent with the values and mission of the CTBUH.

Note: Awards in some categories may not be conferred on an annual basis if the criteria cannot be clearly met or demonstrated through the submittal.

Best Tall Building

Americas

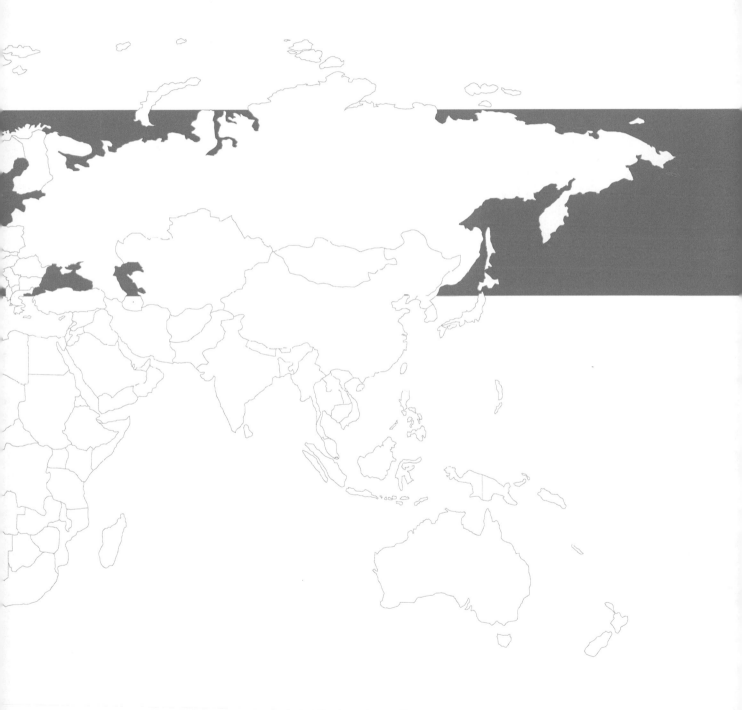

Edith Green-Wendell Wyatt Federal Building

Portland, United States of America

Completion Date: Original: 1974; Renovation: May 2013
Height: 110 m (361 ft)
Stories: 18
Area: 48,774 sq m (524,999 sq ft)
Use: Office
Owner: General Services Administration
Architect: Cutler Anderson Architects (design);
SERA Architects (architect of record)
Structural Engineer: KPFF Consulting Engineers
MEP Engineer: Interface; PAE Consulting Engineers; Stantec
Main Contractor: Howard S. Wright Construction
Other Consultants: Acoustic Design Studio (acoustics); Charles M. Salter
Associates (acoustics); PLACE (landscape)

"A significant transformation both from a performance and urban perspective, this renovated federal building demonstrates how buildings need not be destroyed to gain new life."

Jeanne Gang, Jury Chair, Studio Gang Architects

The Edith Green-Wendell Wyatt (EGWW) Federal Building is an existing 18-story, 48,774 square-meter office tower, completed in 1974. The building no longer met the functional or the energy and conservation requirements of the contemporary US government, so a major renovation project was undertaken. A mechanical upgrade, seismic retrofit, and full interior rehabilitation was paired with a full replacement of the building envelope with a distinctive shading façade, affording better energy performance and a new lease on life.

While investigating the brief, the architects discovered that the existing concrete skin of the structure used up 600 millimeters of floor area for every 300 lineal millimeters of exterior wall. By applying a new skin to the existing slab edges and making other changes related to HVAC systems, the design added 9,449 square meters of new rentable office space, which amortized the added cost of the envelope. About 650 square meters of that space was freed up by switching to water cooling, which reduced the building's thermal load to the point

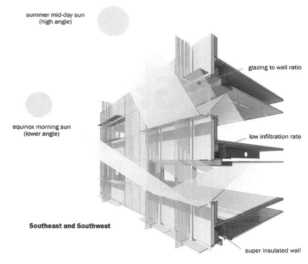

summer mid-day sun
(high angle)

glazing to wall ratio

equinox morning sun
(lower angle)

low infiltration rate

Southeast and Southwest

super insulated wall

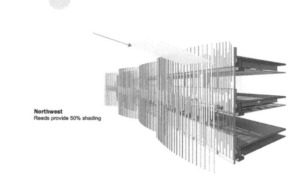

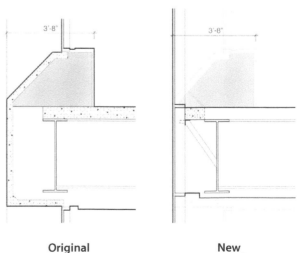

Northwest
Reeds provide 50% shading

that the large fans for the previous forced-air cooling system could be removed. Additionally, removing the concrete façade panels simplified the seismic retrofit of the building due to the reduction of weight; the new façade system could be attached to the structural frame with a series of relatively short steel beams.

Because of the importance daylighting plays in human health and comfort, the project optimized solar penetration in the perimeter zone by deploying a low-powered ambient lighting in concert with focused task lamps. This resulted in a 50–60 percent reduction in energy consumption for lighting, while providing occupants with a valuable connection to the outdoors. The depth and spacing of the shading devices were varied to arrive at the performance metrics the designers used, and to derive the building's aesthetic expression.

In order to respond to the unique solar exposure of the site, each face of the building was designed to both

3'-8"

3'-8"

Original

New

shade direct solar gain and reflect light into the interior spaces to enhance daylighting. The result was a building that presents a different face to each solar circumstance. The reed-like shades affixed to the northwest façade of the building are tuned to reduce solar gain, and a 3,962 square-meter roof canopy supports a 180 kW photovoltaic array, while also collecting rainwater.

Greater than 65 percent water savings will be achieved through a dual strategy of incorporating water-conserving plumbing fixtures together with the rainwater system. The water conservation strategy started with an analysis of how the existing building used water. Eighty-seven percent of the building's water usage is for domestic uses, with 13 percent used for irrigation of surrounding vegetation. Because of this large interior use, the strategy focused on reusing rainwater for non-potable flush fixture uses first. Landscape water use is reduced by over 50 percent as well, through use of drought-resistant landscaping and incorporation of subsurface irrigation.

A 624,593-liter tank, created by repurposing an old rifle range, allows rainwater to be stored and used for toilet flushing, irrigation, and mechanical cooling tower makeup water. The tank also supports another project goal: mitigating the negative effects of urban runoff. Ultimately, the EGWW building is expected to save over 7.5 million liters of water annually – enough water to fill 22 swimming pools. The building is also designed to achieve a 60 percent reduction in energy use compared

Previous Spread

Left: Overall view of the tower and northwest façade

Right: Original view of building before renovation

Current Spread

Opposite Left: View of the southwest façade shading system

Opposite Right: Drawings showing the sun-shading features of the façade (top and middle) and how the additional square footage was achieved (bottom)

Top Left: Typical floor plan

Top Right: View of building and façade shading system from north

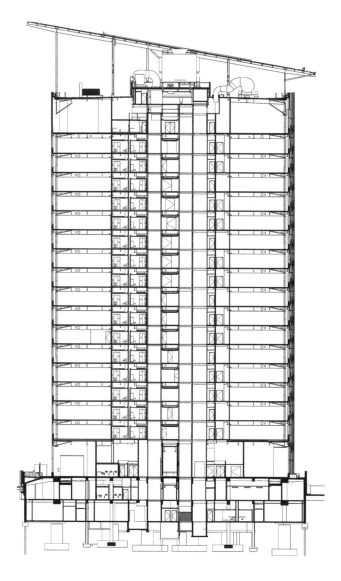

to the existing building, and a reduction of Energy Use Intensity (EUI) by 45 percent. A large portion of those savings will come from having eliminated forced-air fans.

In the original refurbishment design for the building, a green wall system was to be utilized to shade the west façade. Planters were planned vertically across the façade every two floors to create bands of greenery across the entire height of the building. However, it was later determined too costly to implement, and the planters were replaced with the reed-like fixed aluminum shading devices that now characterize the northwest façade. The "green wall" was thus limited to only the first two floors of the building, where it is able to grow out of a planter bed at ground level and use the first levels of the reed shading devices as a trellis to grow upward.

"Even though the green wall was not realized as originally intended, it is remarkable that this building with its 'futuristic' aesthetic had a former life – as an uninspired concrete box. The Edith Green-Wendell Wyatt Federal Building points the way forward on what really can be achieved with the refurbishment of existing buildings."

Antony Wood, Juror, CTBUH

Jury Statement

The renovation of the Edith Green-Wendell Wyatt Federal Building is more than an improvement in energy performance. It is a public restatement of the contract between a government, its people, and the natural environment. Edith Green-Wendell Wyatt is a retrofit of a building that was designed at the time of the Arabian Oil Embargo and in the shadow of Watergate, and all of the attendant paranoia and utilitarianism of civic architecture at the time.

It now emerges at a time when energy is again the focus, but it underscores just how much the attitude toward energy, as well as cities, the environment, and workplace design have changed, even as requirements for physical and information security have increased. Given that the Edith Green-Wendell Wyatt building communicates openness, urban vitality, and sustainability, yet still meets stringent operational requirements, it is all the more remarkable that the original met none of these criteria. The fact that it still seems like an "outlier" when we think of "government building in America" tells us how much work we still have to do, especially with our existing building stock.

The Point
Guayaquil, Ecuador

Completion Date: March 2014
Height: 137 m (449 ft)
Stories: 36
Area: 130,000 sq m (1,399,308 sq ft)
Use: Office
Owner/Developer: Pronobis
Architect: Christian Wiese Architects
Structural Engineer: Ernesto NA
MEP Engineer: Ernesto NA
Main Contractor: Inmo Mariuxi
Other Consultants: Adriana Hoyos (interiors); Coheco (vertical transportation); Consuambiente (environmental); Coyado (lighting); Imecanic (fire)

"Sited on the riverbank, the tower captures the waterfront landscape through its sculptural form, which effectively resembles wave movement through its rotating floor plates."

Wai Ming Tsang, Juror, Ping An Development

The Point is the tallest building in Ecuador, at 137 meters. It takes the opportunity presented by its prominent place in the skyline to experiment with the traditional skyscraper form, by undulating as a sculpture, stepping out of the way of key views, while becoming a key view in and of itself.

The site of the building is significant: the confluence of the Babahayo and Daule rivers forms the Guayas River, which connects Guayaquil to the Pacific Ocean and serves as a deepwater port for Ecuador. The swirling shape of the tower is meant to represent the whirlpools that occur at the confluence. The establishment of The Point and its surrounding 130,000 square-meter Ciudad del Rio project are acknowledgements of Guayaquil's rediscovery of its river. The Point and Ciudad del Rio were conceived as a sustainable development, one in which the buildings occupy as little area as possible, freeing up space for pedestrians strolling along the riverside.

> *"With The Point, Ecuador now has it's feminine, graceful flowing 'organic' tower, like Aqua in Chicago, Absolute in Mississaugua, and others..."*
>
> Antony Wood, Juror, CTBUH

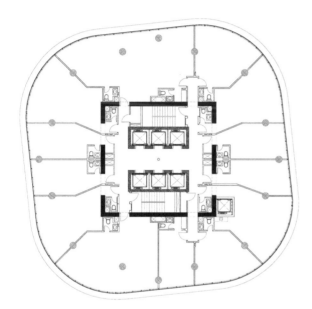

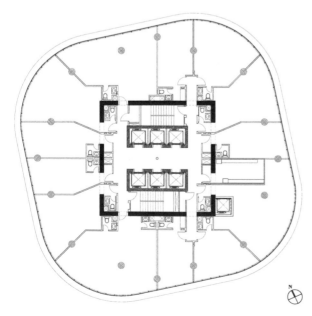

In some ways, The Point and Ciudad del Rio, beginning construction in 2011, are capstones of a decades-long resurgence for Guayaquil, beginning in the 1990s. In an effort to regenerate the urban area, power and telephone lines were buried underground, the Guayas River promenade was restored with a new boardwalk in 2000, and a bus-rapid-transit system was implemented in 2006, making the city friendlier to business and tourism and enhancing the quality of life for citizens.

The program for the 35-story Point consists of offices on the upper 33 floors and an executive club on the two highest floors, with the ground floor dedicated to commercial use. The executive club contains amenities, such as a solarium with panoramic view, Jacuzzi, sauna, recreational area, bar/lounge, gym and restaurant. The basic floor plan of the building is a square, with its edge rounded to generate the shape of the building through a 6-degree rotation between floors, the shape echoes the flowing water of the Guayas River. The rounded corners cut

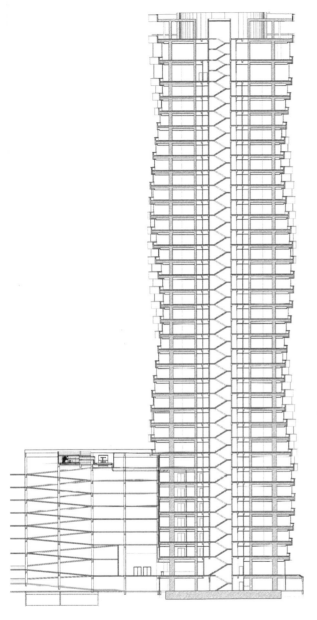

the necessary length of the concrete cantilevers, permitting their design to be simple and efficient, and their construction to be carried out largely by local Ecuadorian labor. Most of the materials used in the building were produced in Ecuador and/or within an 800 kilometer radius of the project.

LEED principles guided many of the design choices, and were followed throughout the construction process. The cantilevered design affords some solar shading, lowering interior temperatures. Rainwater is captured for flushing bathroom fixtures and for irrigation of the grounds, which are planted with native species. The building's glazing consists of thermally efficient glass. LED lighting is used throughout the interior, while exterior lighting is solar-powered.

The curving, cantilevered form has already won hearts and minds in Ecuador, where its model can be found in many stores as a souvenir of Guayaquil. It even has a nickname: "the screw."

Jury Statement

With The Point, Ecuador joins the growing ranks of South American nations that recognize the significance of a definitive icon for their cities. Importantly, in this case, the icon also serves as a catalyst for urban regeneration. In addition to attracting commercial and tourism activity to the center of town, The Point also reintroduces the citizens of Guayaquil to the river that made their city. That the tower was designed and constructed largely by local forces also speaks to the commitment Guayaquil has made to itself.

Previous Spread

Left: Overall view of tower from the boardwalk

Right: Detail of façade representative of flowing water

Current Spread

Opposite Left: Typical floor plans with six degree rotation per floor

Opposite Right: Overall view of the tower

Left: Building section

Right: Tower detail of the façade and balconies

United Nations Secretariat Building

New York City, United States of America

Completion Date: Original: 1951; Renovation: September 2013
Height: 154 m (506 ft)
Stories: 39
Area: 81,940 sq m (882,000 sq ft)
Use: Office
Owner/Developer: United Nations
Architect: HLW International LLP (interiors); R.A. Heintges & Associates (façade)
Structural Engineer: HLW International LLP
MEP Engineer: SYSKA Hennessy Group
Project Manager: Gardiner & Theobald Inc
Main Contractor: Skanska
Other Consultants: di Domenico + Partners, LLP (landscape); Kroll Inc (security); Rolf Jensen & Associates (fire); Shen Milsom Wilke, Inc. (acoustics); Viridian Energy & Environmental, LLC (sustainability); Weidlinger Associates (security)

"A perfect example of a rejuvenation of a building where the increase in the performance goes hand in hand with respect for the original architectural intent."

David Gianotten, Juror, OMA

The 39-story glass-and-steel Secretariat tower of 1951 is arguably the most visible representation of postwar optimism and resiliency on the United Nations campus in New York. It also embodies a mid-century Modernist merging of technology and form, as expressed in the remarkably slim north-south facing profiles and the crystalline east-west elevations.

In 2008, when the six-year, multi-billion dollar renovation of the UN campus started, the Secretariat, which has never ceased to operate as a functional government building throughout the refurbishment, was plagued by severely outdated fixtures, deficient life safety features, and a leaky curtain wall. It is estimated that the historic building had operated at least 35 years beyond its normal lifespan.

Key design strategies for increasing performance included the extensive redevelopment of the base building core and systems, including new elevator systems and mechanical infrastructure, new fire

protection systems and code upgrades, disabled-access compliance, and asbestos abatement. Floor plate utilization was enhanced through the introduction of a new planning diagram. It also involved installing a better, more efficient envelope for insulation and blast protection.

The Secretariat was an early example of a tall building to employ a suspended wall system. Its primary elevations were enclosed by free-hanging

"I really like the sensitive restoration which recycles for longer life a much loved icon. Its setting and its simplicity make what could be a universal shape quite singular, quite unique."

Sir Terry Farrell, Juror, Farrells

glazed façades. The new curtain wall was designed to look like it did in 1951 – sleek and taut, with its double-hung aluminum windows, glazed spandrel panels, and aluminum-clad steel mullions appearing as one continuous transparent form. The replaced wall's appearance had been significantly altered after nearly 60 years of patches, caulkings, insulating and blast coatings, resulting in a patchwork appearance and a greenish hue.

To rectify this, the project team conducted extensive testing to replicate the visual appearance of the original building, including reflection patterns at different times of day. The work included spectral analysis of glass types to identify viable formulas, computer model simulation, and, ultimately, a full-scale mockup tested on the UN grounds.

The original curtain wall was demolished and replaced with a pressure-equalized system in sections, proceeding from bottom to top in each 10-story zone between the louvered mechanical levels. Unlike the original curtain wall, which had been attached to the concrete

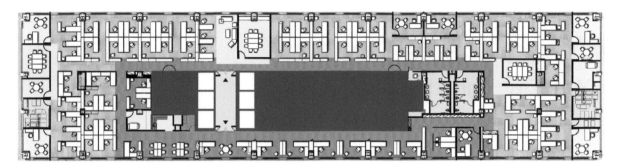

floor slabs, the new system is connected to the now-reinforced building frame via outrigger plates. To mimic the appearance of the original double-hung windows, the team strategically offset some of the aluminum extrusions. The new cladding incorporates performance enhancements, such as low-E coating and blast protection.

The new open floor plan challenges the closed culture propagated by the original cellular offices. Significantly more light and air now enter the interior. The open plan also affords more flexibility and efficient use of space in a rapidly growing organization; the Secretariat now serves 193 member-states, as compared to the original 50.

The open-plan daylighting scheme includes 1.9-meter-tall furniture-work walls that extend from perimeter columns, without blocking natural light. The ceiling is 2.4 meters high at center of plan but gradually steps up to 2.9 meters at the windows, allowing for a circulation path around the core, while supporting the equitable

distribution of mechanical services. Seventy-five percent of workspaces now have daylighting and views.

Overall, the resulting building is 50 percent more energy-efficient than it was before the renovation. Total building replacement was never an option due to the iconic nature of the tower; instead, the classic building has been thoroughly renovated in place. In many ways, the Secretariat's radical revitalization is more about rebirth than restoration.

Jury Statement

The design team's careful consideration and planning has been able to breathe new life into this iconic UN building. The performance enhancements and open floor plans have transformed the building into a model workplace worthy of its owner's stature in the world. It was important that this icon of mid-century optimism could transcend the limitations of the technology of the time, and the formidable growth in its constituency, to once again broadcast the United Nations mission in the best possible light.

4 World Trade Center
New York City, United States of America

4 World Trade Center is the first tall building to open on the actual 16-acre World Trade Center rebuilding site. The glass-enclosed tower allows for a play of light along its surface, such that the substantial mass of the building sometimes seems to disappear against the sky. A large outdoor terrace is located on the roof of the setback, offering a spacious open-air space, an unusual feature for a Manhattan office building.

The building is as transparent to the memorial site across the street as it is deeply interwoven with the elaborate network of tunnels connecting the World Trade Center to New York's subway system. Visitors arriving to the building lobby from below are greeted by the serene monolith of black Swedish granite, offset by 14-meter floor-to-ceiling windows that preview the green oasis of the memorial park. The building's commitment to sustainability includes 100 percent renewable energy sourcing, and rainwater collection for cooling tower replenishment.

Completion Date: November 2013
Height: 298 m (978 ft)
Stories: 65
Area: 232,258 sq m (2,500,004 sq ft)
Use: Office
Owner/Developer: Silverstein Properties
Architect: Maki and Associates (design); Adamson Associates (architect of record)
Structural Engineer: Leslie E. Robertson Associates
MEP Engineer: Jaros Baum & Bolles Consulting Engineers
Main Contractor: Tishman Construction
Other Consultants: Code Consultants Professional Engineers, PC (code); Ducibella Venter & Santore (security); S.D. Keppler & Associates, LLC (LEED)

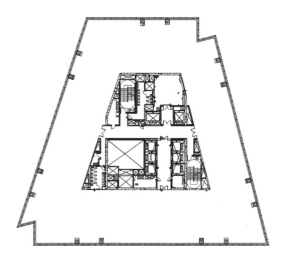

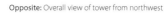

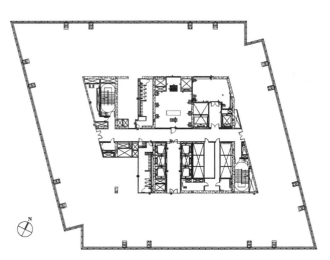

Opposite: Overall view of tower from northwest

Above: Typical floor plans – levels 58 to 61 (top) and levels 44 to 50 (bottom)

Top Right: Context view from Church Street

Middle Right: Interior space

Bottom Right: Entrance approach with water feature

Nominee
Best Tall Building Americas

Magma Towers
Monterrey, Mexico

Magma Towers is a residential project located in the exclusive area of Valle Oriente, Monterrey, Mexico, consisting of two towers containing 230 apartments, mounted over a two-level commercial base. The two dark grey towers have a somewhat monolithic appearance, which is enlivened by the faceted geometry of the façade. The space that separates the two towers at the lobby level contains most of the communal amenities, such as the swimming pool, bar, gym, green roof garden, as well as other common areas.

The rhythm of the façade pattern is randomized, but corralled so that it could fit economically into the suspended façade system, and adjusted to suit the wide variations in floor plan demanded by the program. The subtractions in the façade generate balconies for the apartments they adjoin. Inside, the program is highly varied among housing configurations, with some units having balconies, some without. Double-height lofts are also interlocked with standard one- and two-bedroom apartments.

Completion Date: July 2013
Height: West Tower: 105 m (344 ft); East Tower: 85.5 m (281 ft)
Stories: West Tower: 28; East Tower: 24
Total Area: 27,437 sq m (295,329 sq ft)
Use: Residential
Owner/Developer: Grupo Inmobiliario Monterrey
Architect: GLR Arquitectos
Structural Engineer: Socsa
Main Contractor: Anahuac Organización Constructora
Other Consultants: Harari LA (landscape)

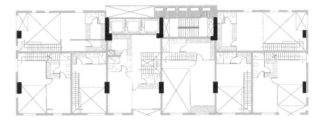

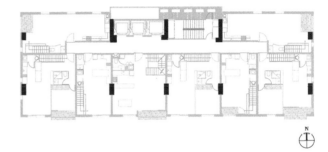

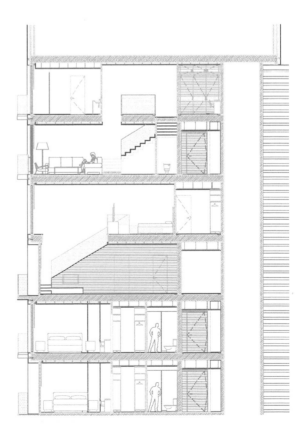

Opposite: Overall view of tower from southeast

Top Left: View of tower and retail from street level

Bottom Left: Façade detail

Top Right: Typical floor plans – upper loft level (top) and lower loft level (bottom)

Bottom Right: Detailed Section

MuseumHouse
Toronto, Canada

MuseumHouse, a 19-story glass and steel residential building with a two-story retail/amenity base, is conceived as a simple, elegant and complementary counterpoint to Daniel Libeskind's iconic and complex Crystal at the Royal Ontario Museum across the street. Sited on a narrow infill property 40 meters deep, 13 meters wide on one side and 16 meters wide on the other, MuseumHouse contains only one or two residences per floor. Each has direct access elevators and north- and south-facing limestone loggias and stainless-steel planter boxes. Space is further economized by a five-level automated parking system below grade.

All units were designed to take advantage of spectacular views with generous terraces to the north and south. The building is clad in local materials such as Ontario limestone, metal, and curtain-wall glass with high-performance coatings. The continuous planter boxes are automatically irrigated and drained, enriching the building façade with greenery and augmenting the shade of the loggias.

Completion Date: July 2013
Height: 75 m (246 ft)
Stories: 19
Area: 7,751 sq m (83,431 sq ft)
Use: Residential
Owner: MuseumHouse Condominium Corporation
Developer: 206 Bloor Street West Development Corporation
Architect: Page + Steele / IBI Group Architects
Structural Engineer: Jablonsky Ast & Partners
MEP Engineer: Merber Corporation
Project Manager: Yorkville Construction Corporation
Main Contractor: Veisman Consulting Limited
Other Consultants: Land Art Design (landscape)

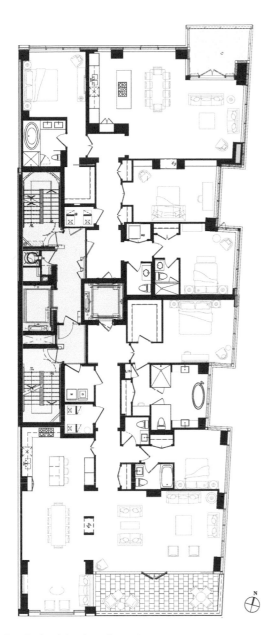

Opposite: Overall view of tower from south

Above: Typical residential floor plan

Top Right: Looking up at "green" façade created by continuous planter boxes on balconies

Bottom Right: Overall view with part of Libeskind's "Crystal" in the foreground

Peninsula Tower
Mexico City, Mexico

The Peninsula Tower is a residential skyscraper located in Santa Fe, an urban development west of Mexico City. The building is part of a real estate project of three contiguous plots of land who will build three apartment towers joined by a commercial area, in separate stages. The Peninsula Tower marks the completion of stage one. There are 39 apartment floors, a two-level penthouse, and a roof garden and heliport. Three additional floors house amenities such as a pool, spa, gym, and projection room.

While basically a rectilinear building, the tower breaks up its mass by way of diagonals set in different positions on each of the four façades, creating a dynamic effect that accentuates the tower's verticality. The glazed envelope of the building is set back 1.27 meters from the surface of the chiseled white concrete perimeter columns, providing shadow interplay for exterior visual interest and solar glare protection for the interior.

Completion Date: September 2014
Height: 164 m (538 ft)
Stories: 50
Area: 45,577 sq m (490587 sq ft)
Use: Residential
Owner/Developer: Residencial Peninsula Santa Fe
Architect: Teodoro González de León
Structural Engineer: Luis Bozzo Estructuras y Proyectos S.L.
MEP Engineer: Ingenieria y Controles Coyoacán; Instalaciones Planificadas S.A. de C.V.; Instalaciones de Aire S.A. de C.V.
Project Manager: Citicapital
Main Contractor: Anteus Constructora
Other Consultants: 4 Estaciones (landscape); Desarrollo Aluminero Lea (façade); JOOFarrill M Arquitectos S.A. de C.V. (façade)

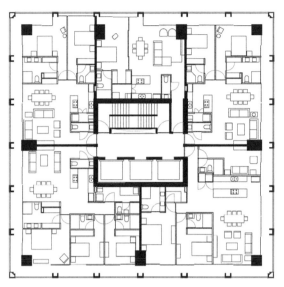

Opposite: Overall view of tower from southeast

Above: View of tower from street level

Top Right: Typical floor plan – levels 10 to 47

Bottom Right: Façade view from tower base

Territoria El Bosque
Santiago, Chile

Territoria El Bosque is a LEED Gold office building in Apoquindo, the business and financial district of Santiago, Chile. Situated at an important intersection, the site is adjacent to a 300-year old ceibo, a flowering tree found throughout South America. The tree has become a local landmark, and its prominence and beauty heavily influenced the building's architectural expression.

A reflective glass covers the northern volume, while the southern volume incorporates a mesh screen into the façade, giving the building texture and distinctly breaking up the mass. The screens, which vary in opacity depending on which façade they sit, are placed slightly away from the building, which allows the light to play on the glass behind them. The building's most notable feature directly references the landmark and Santiago's history of incorporating landscape into building façades: circular tree-boxes contain camphor trees at projecting balconies on every third floor. Additionally, the rooftop garden features extensive landscaping with local greenery, and was designed to be highly water efficient.

Completion Date: August 2013
Height: 74 m (243 ft)
Stories: 20
Area: 32,000 sq m (344,445 sq ft)
Use: Office
Owner/Developer: Territoria
Architect: Handel Architects (design); Uno Proyectos (architect of record)
Structural Engineer: Santolaya Ingenieros Consultores (design); Alonso Larrin Evaristo (peer review)
MEP Engineer: Gormaz y Zenteno, Ltda.; Fleischmann Ingenieria de Proyectos, Ltda.; Enrique Montoya Ingenieria, Ltda.
Main Contractor: Constructora Sigro S.A.
Other Consultants: Accura Systems (façade); AEV Topografia (civil); EBV Fire & Security (life safety); Energy 3Arq (LEED); Heavenward (vertical transportation); Ruz & Vukasovic (geotechnical)

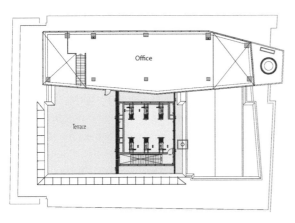

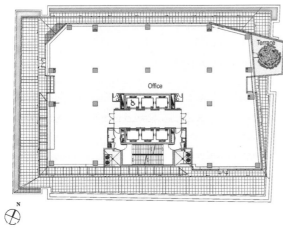

Opposite: Overall view of tower from northeast featuring cantilevered tree balconies

Above: Floor plans – roof terrace level (top) and typical office level (bottom)

Top Right: View of external mesh façade screen from south

Bottom Right: Public space on ground level

The Godfrey
Chicago, United States of America

The Godfrey Hotel stands as a kind of redemption of an ambitious, but stalled project, resulting in an improvement on the original concept. The original project intended to accommodate a low-amenity extended-stay suite hotel, began construction in 2007 but was halted by the economic downturn of 2008. The partially constructed structure, wrapped in a tarp, was nicknamed "The Mummy" and began to rust. Fortunately, the unusual staggered steel truss frame design drew the attention of a new developer, which in 2011 engaged the original architect, to finish the project as a "boutique/lifestyle" hotel.

The 16-story, metal-skinned, shifted-box structure was reclad, and the interiors adjusted to fully embrace the structurally expressive design. Guest room views are occasionally interrupted by a plunging K-brace, which may not be to everyone's taste, but will appeal to aficionados of Modernism and Chicago's brawny industrial history. Solace also comes in opportunistic exploitation of the shifted-box parti, in the form of a 1,000 square-meter rooftop space under a retractable glass enclosure.

Completion Date: January 2014
Height: 56 m (184 ft)
Stories: 16
Area: 16,410 sq m (176,636 sq ft)
Use: Hotel
Owner/Developer: Oxford Capital Group, LLC
Architect: Valerio Dewalt Train Associates, Inc.
Structural Engineer: Structural Affiliates International, Inc.
MEP Engineer: WMA Consulting Engineers, Ltd.
Project Manager: Daccord
Main Contractor: Lend Lease
Other Consultants: Gettys Group Hospitality Design (interiors); Nova Fire Protection (fire); Schuler Shook (lighting); Shiner + Associates (acoustics); V3 Companies (civil); Wolff Landscape Architecture (landscape)

Opposite: Overall view of tower from northwest

Above: Floor plan – typical hotel room level

Top Right: West façade showing detail of K-brace expression

Bottom Right: Hotel room with distinctive exposed K-brace

Torre Costanera
Santiago, Chile

At 300 meters, Torre Costanera is the tallest building in South America. Santiago's close proximity to the Andes, and the need to distinguish the tower against this dramatic backdrop, has prompted a simple and clear form. Rising from the northwest corner of the development next to the Mapocho River, the glass-clad tower has a slightly tapered, slender form that culminates in a sculptural latticed crown. This dramatic steel and glass structure, rising 45 meters in height, provides a unique and elegant silhouette. The four corners are indented to accentuate its slenderness. The glass surface of the tower strikes a delicate balance between transparency and reflectivity.

This is a 21st-century building, both technically and aesthetically. It is designed with state-of-the-art structural and mechanical systems, including a highly advanced outrigger system to account for Santiago's high level of seismic activity. The cooling tower draws its entire water supply from the adjacent canal.

Completion Date: August 2014
Height: 300 m (984 ft)
Stories: 62
Area: 110,000 sq m (1,184,030 sq ft)
Use: Office
Owner/Developer: Cencosud S.A.
Architect: Pelli Clarke Pelli Architects (design); Alemparte, Barreda y Asociados (architect of record)
Structural Engineer: René Lagos y Asociados; Thornton Tomasetti
MEP Engineer: Inspecta
Main Contractor: Salfa Corporation
Other Consultants: ALT Cladding (façade); RWDI (wind); Watt International (energy concept)

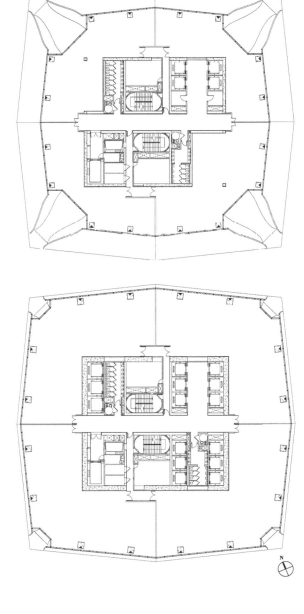

Opposite: Overall view of tower from northeast

Top: Overall view of tower in skyline context

Left: View of tower as from street

Right: Floor plans – high-rise (top) and mid-rise level (bottom)

500 Lake Shore Drive
Chicago, United States of America

Completion Date: May 2013
Height: 152 m (497 ft)
Stories: 47
Area: 66,363 sq m (714,321 sq ft)
Use: Residential
Owner/Developer: Related Midwest
Architect: Solomon Cordwell Buenz
Structural Engineer: Magnusson Klemencic Associates
MEP Engineer: Cosentini
Main Contractor: Lend Lease
Other Consultants: Engineering PLUS, LLC (security); dbHMS (LEED); Jenkins & Huntington, Inc. (vertical transportation); Oslund & Associates (landscape); Shiner + Associates (acoustics); Robert Pope Associates, Inc. (interiors); W. J. Higgins & Associates, Inc. (façade)

1812 North Moore Street
Arlington, United States of America

Completion Date: November 2013
Height: 118 m (387 ft)
Stories: 30
Area: 53,883 sq m (579,992 sq ft)
Use: Office
Owner/Developer: Monday Properties
Architect: Davis, Carter, Scott, Ltd.
Structural Engineer: KCE Structural Engineers
MEP Engineer: Dewberry (formerly TOLK)
Main Contractor: Clark Construction
Other Consultants: MCLA Lighting Design (lighting); Sustainable Design Consulting, LLC (LEED); Van Deusen & Associates (vertical transportation); VIKA, Inc. (civil); RWDI (wind)

Concord Cityplace Parade

Toronto, Canada

Completion Date: 2013
Height: Tower S: 132 m (432 ft); Tower V: 116m (381 ft)
Stories: Tower S: 45; Tower V: 40
Use: Residential
Owner/Developer: Concord Adex
Architect: Kohn Pedersen Fox Associates (design); Page + Steele / IBI Group Architects (architect of record)
Structural Engineer: Yolles, a CH2M HILL Company
MEP Engineer: MCW Consultants, Ltd.
Main Contractor: PCL
Other Consultants: BVDA (façade); Ferris+Associates, Inc. (landscape); Golder Associates, Ltd. (acoustics); II By IV Design Associates, Inc. (interiors); Larden Muniak Consulting, Inc. (code); Lea Consulting, Ltd. (traffic); Mike Niven Interior Design (interiors); MMM Group (civil); RWDI (wind)

Courtyard & Residence Inn Manhattan/Central Park

New York City, United States of America

Completion Date: December 2013
Height: 230 m (755 ft)
Stories: 67
Area: 29,058 sq m (312,778 sq ft)
Use: Hotel
Owner/Developer: Granite Broadway Development, LLC
Architect: Nobutaka Ashihara Architect, PC
Structural Engineer: WSP Group
MEP Engineer: Edward & Zuck
Main Contractor: GCNY Group
Other Consultants: Alan G. Davenport Wind Engineering Group BLWTL (wind); Bill Rooney Studio, Inc. (interiors); Israel Berger & Associates (façade); Jenkins & Huntington, Inc. (vertical transportation); Langan Engineering & Environmental Services (geotechnical); Shen Milsom Wilke, Inc. (acoustics)

Couture
Toronto, Canada

Completion Date: August 2013
Height: 137 m (449 ft)
Stories: 42
Area: 36,490 sq m (392,775 sq ft)
Use: Residential
Owner/Developer: Monarch Couture Developments, Ltd.
Architect: Graziani + Croazza Architects
Structural Engineer: Reed Jones Christoffersen Consulting Engineers
MEP Engineer: MV Shore Associates, Ltd.
Main Contractor: Monarch Corporation
Other Consultants: Gradient Microclimate Engineering, Inc. (environmental); HGC Engineers (acoustics); MBTW (landscape); McClymont + RAK Engineers, Inc. (geotechnical); Mike Niven Interior Design (interiors); Stantec (servicing engineer); R.Avis Surveying, Inc. (quantity surveyor)

K2 at K Station
Chicago, United States of America

Completion Date: March 2013
Height: 112 m (367 ft)
Stories: 33
Area: 56,671 sq m (610,002 sq ft)
Use: Residential
Owner/Developer: Fifield Companies; Wood Partners
Architect: Pappageorge Haymes Partners
Structural Engineer: Samartano & Company
MEP Engineer: Advance Mechanical Systems, Inc.; Warren Thomas Plumbing Co.; Chatfield Electric, Inc.
Main Contractor: McHugh Construction
Other Consultants: Charter Sills & Associates (lighting); Eriksson Engineering Associates, Ltd. (civil); GSG Consultants, Inc. (geotechnical); Hollingsworth Architects, LLC (façade); Jenkins & Huntington, Inc. (vertical transportation); Shiner + Associates (acoustics); Adrian Smith + Gordon Gill Architecture (interior); Terra Engineering, Ltd. (landscape)

NEMA
San Francisco, United States of America

Completion Date: March 2014
Height: 118 m (387 ft)
Stories: 37
Area: 184,404 sq m (605,000 sq ft)
Use: Residential
Owner/Developer: Tenth and Market, LLC
Architect: Handel Architects
Structural Engineer: Magnusson Klemencic Associates
MEP Engineer: CB Engineers
Main Contractor: Swinerton Builders

The John and Frances Angelos Law Center
Baltimore, United States of America

Completion Date: April 2013
Height: 71 m (231 ft)
Stories: 12
Area: 17,645 sq m (189,929 sq ft)
Use: Education
Owner: University of Baltimore
Architect: Behnisch Architekten (design); Ayers Saint Gross (architect of record)
Structural Engineer: Cagley & Associates
MEP Engineer: Mueller Associates; Diversified Engineering
Main Contractor: The Whiting-Turner Contracting Company
Other Consultants: Aon Fire Protection Engineering Corporation (fire); AECOM / Davis Langdon (cost); MCLA Lighting Design (lighting); Shen Milsom Wilke, Inc. (acoustics); Stephen Stimson Associates (landscape); Transsolar (energy concept); RK & K (civil); Rolf Jensen & Associates (code)

The Peter Gilgan Centre for Research and Learning

Toronto, Canada

Completion Date: September 2013
Height: 118 m
Stories: 22
Area: 86,880 sq m (935,168 sq ft)
Use: Education
Owner/Developer: The Hospital for Sick Children
Architect: Diamond Schmitt Architects
Structural Engineer: Yolles, a CH2M HILL Company
MEP Engineer: H.H. Angus & Associates, Ltd.; MMM Group
Main Contractor: EllisDon Corporation
Other Consultants: Aercoustics Engineering (acoustics); BA Group (traffic); CDML Consulting, Ltd. (sustainability); dtah (landscape); HDR Architecture (lab consultant); H.H. Angus & Associates, Ltd. (vertical transportation); Leber / Rubes Inc. (fire); RWDI (wind)

Torres del Yacht

Buenos Aires, Argentina

Completion Date: February 2014
Height: 135 m (443 ft)
Stories: 45
Use: Residential
Owner: Sociedad Inversora Dique IV S.A.
Developer: Fernandez Prieto Desarrollos Inmobiliarios S.A.
Architect: Fernandez Prieto & Asoc. Ingenieros y Arquitectos S.A.; M.SG.S.S.S. Arquitectos
Structural Engineer: Fainstein AHF S.A.
MEP Engineer: Labonia; Estudio Grinberg GF S.A.
Main Contractor: EDINTAR construcciones S.A.; EMEPA; IECSA
Other Consultants: Angeli (façade); CPP (wind); Lucciola – Fass Yakov (lighting)

YooPanama Inspired by Starck

Panama City, Panama

Completion Date: November 2013
Height: 247 m (810 ft)
Stories: 78
Use: Residential
Owner/Developer: Arts Towers Development S.A.
Architect: Bettis Tarazi Arquitectos S.A.
Structural Engineer: OM Ramirez & Associates S.A.
MEP Engineer: Plumbing Corporation, Inc.; Ingenieria Atlantico; Ingeniería Carpenn
Main Contractor: Syasa Panama

ZenCity

Buenos Aires, Argentina

Completion Date: March 2014
Height: 77 m (252 ft)
Stories: 25
Use: Residential
Owner: Fideicomiso Puerto Madero 7
Developer: Fernandez Prieto Desarrollos Inmobiliarios S.A.; Intelligent; Furigo; Servicios Portuarios
Architect: Fernandez Prieto & Asoc. Ingenieros y Arquitectos S.A.; Bodas, Miani, Anger Arquitectos & Asociados
Structural Engineer: AHF.sa Ingenieros Estructurales; Raul A. Curuchet José Maria Del Villar Ingenieros Civiles
MEP Engineer: G.N.B.A. Consultores S.R.L.
Main Contractor: Caputo S.A. & Xapor S.A.
Other Consultants: CPP (wind); Churba / Bernardini (interior); Dakno S.A. (façade); Iluminación Sudamericana S.A. (lighting); Ing. Eugenio Mendiguren (geotechnical)

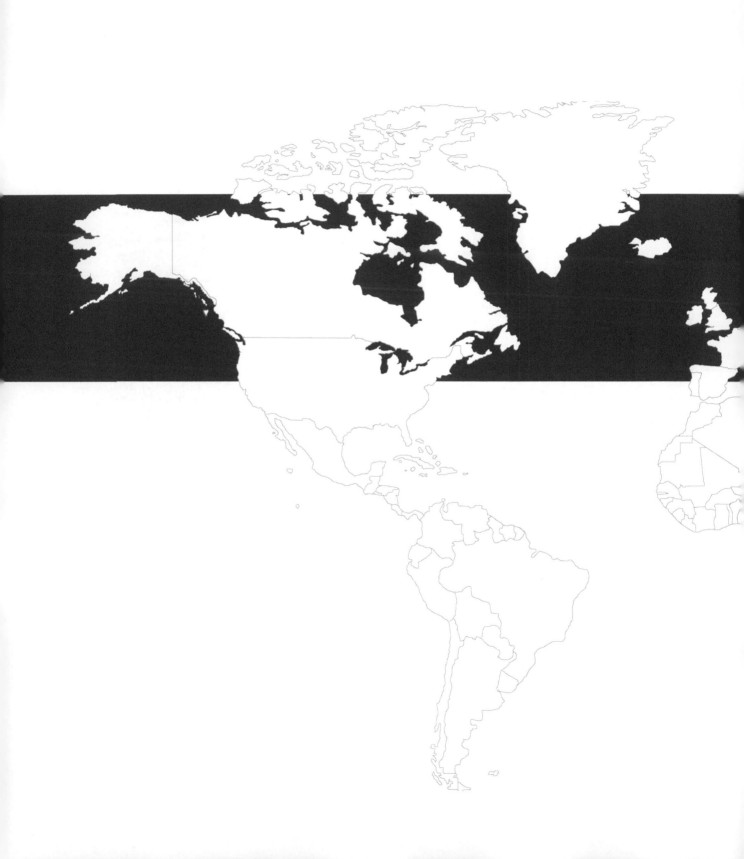

Best Tall Building
Asia & Australasia

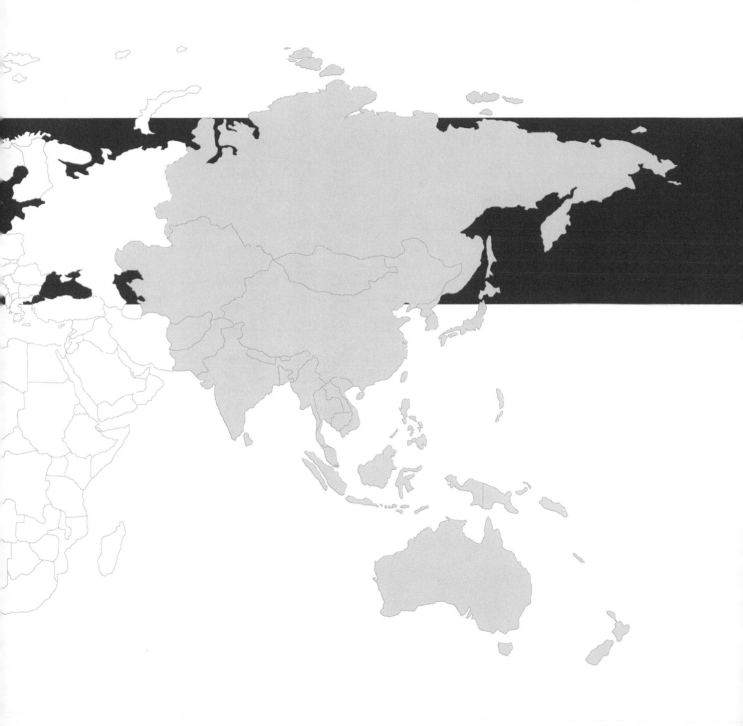

Winner
Best Tall Building Asia & Australasia

One Central Park
Sydney, Australia

Completion Date: January 2014
Height: 116 m (381 ft)
Stories: 34
Area: 67,626 sq m (727,920 sq ft)
Use: Residential
Owner/Developer: Frasers Property Australia; Sekisui House Australia
Architect: Ateliers Jean Nouvel (design); PTW Architects (architect of record)
Structural Engineer: Robert Bird Group
MEP Engineer: Arup
Main Contractor: Watpac Construction
Other Consultants: AECOM / Davis Langdon (cost); AIK (heliostat lighting); Arup (environmental); Aspect Oculus (landscape); Device Logic (heliostat programming); Jean-Claude HARDY (landscape); Jeppe Aagaard Andersen (landscape); Kennovations (heliostat design); Patrick Blanc (green walls); Surface Design Pty Ltd (façade); Transsolar (Energy); Turf Design (landscape)

"The ubiquitous use of organic shading is designed to improve energy performance and will bring delight to the occupants and its neighbors."

David Scott, Technical Jury Chair, Laing O'Rourke

One Central Park (OCP) is an innovative and environmentally ambitious landmark project within the redevelopment of the Carlton & United Brewery site near Central Station in Sydney. The overall planning intent is to adhere to the highest standards of sustainable residential design under the Australian Green Star rating system and support the vision of an environmentally responsible future for the city.

In order to make the two towers of OCP visibly greener than is normally perceivable in Green Star developments, the design takes a broader approach to carbon-conscious design. With the help of two unusual technologies – hydroponics and heliostats – plants are grown all around the building to provide organic shading, and direct sunlight is harvested all year long for heating and lighting. The shading saves cooling energy, while the redirected sunlight is an all-year light source for the building precinct and adjoining park. Beyond the bravado of their technical deployment and performance, the plants and reflected daylight are also just natural

Previous Spread

Left: Overall view of tower from northwest

Right: Planters along façade create a natural texture

Current Spread

Left: Detail view of the vertical green wall "ribbons" alongside the horizontal planters

Right: Aerial view of building

Opposite Top: View looking up at the heliostat reflectors hanging from the cantilevered sky garden

resources, made available in an unusual way for the enjoyment of Sydney's residents.

Of particular note is the inclusion of a park at the tower's base. The first design challenge was to give the new park a real presence at an urban scale. Because OCP is a high-rise, it is possible to bring the park up into the sky along its façades and make it visible in the city at a distance. On the south side, the park rises in a sequence of planted plateaus scattered like puzzle pieces in randomized patterns across the façades, so that each apartment has not only a balcony, but also its own piece of the park. At the individual scale, this creates pleasant private gardens. At a collective scale, it's a green urban sculpture.

On the north, east and west sides, the green takes more continuous veil-like appearances with green walls, continuous planter bands and climbing vegetation. The plants deliver a message of sustainability, and because their shade reduces energy consumption for

cooling and their leaves sequester carbon dioxide, they also effectively make the building more sustainable. The plants also reflect less heat back into the city than traditional fixed shading. The plants are irrigated with recycled grey and black water, and their growth can be custom-tailored to the needs of each façade area. In total, more than 5 kilometers of planters function like permanent shading shelves and reduce thermal impact in the apartments by up to 30 percent.

A design challenge arises from the tall massing along the north side of the site. In order to remediate overshadowing of the park, the volume is broken up into a lower and a taller tower. On the roof of the lower tower, 42 heliostats (sunlight tracking mirrors) redirect sunlight up to 320 reflectors on a cantilever off the taller tower, which then beam the light down into areas that would otherwise be in permanent shade. The system adapts hourly and seasonally to the need for brightness and warmth, redirecting sunlight to a heat absorbing pool of water atop the atrium glass in summer, which

Jury Statement

One Central Park is a breathtakingly beautiful building that captured the imagination of the jury. The living façades in One Central Park provide fantastic visual, tactile, aromatic, and auditory experiences for the occupants of the apartments and deliver significant urban heat island reductions and other benefits to the local neighborhood. This is also a tall building that welcomes the sun and treats it as an asset to be managed. In addition to shading itself and increasing its own value, the neighborhood is further enhanced by the project's 42 heliostats that reflect sunlight onto the shaded streets. It is perhaps this generosity toward the urban realm that will endure as the project's greatest sustainable achievement.

One Central Park is a truly green building that convincingly shows that tall buildings can be environmentally sound. This is a massive commitment to organic cladding and is commended. It is now hoped that in the coming years the project will be able to demonstrate its intended energy performance, by submitting for the CTBUH Performance Award.

can be drained to assist with heating in the winter. The system redirects up to 200 square meters of direct sunlight and utilizes approximately 40 percent of the corresponding power during Sydney's 2,600 annual sunshine hours. The performance of this system is not accounted for in the BASIX (Building Sustainability Index of New South Wales) and Green Star calculations. With the plants and heliostats, the building thus exceeds its 5 Star rating.

During the day, dappled lights move on the ground in a precisely programmed choreography. At night, the heliostat becomes a monumental urban chandelier and appears in the dark sky like a floating pool of tiny LED lights that merge into a giant screen and simulate reflections of glittering harbor waters.

One Central Park assists Sydney in its goal of moving towards improving its carbon footprint. The project provides much-needed apartments in the city's main job market at its urban core. It reduces energy consumption by 25 percent when compared to a conventional building of its size. The thermal impact of sunlight on the building is reduced by up to 30 percent because of the presence of greenery.

CTBUH Innovation Award Finalist

In addition to being recognized as the winner of the Best Tall Building Asia & Australasia region, the extensive application of greenery has also been recognized by the Awards' technical jury as an Innovation Award Finalist.

Above: View of the cantilevered sky garden at night

Opposite: Level 29 floor plan with cantilevered sky garden and heliostat structure (top) and level 6 floor plan (bottom)

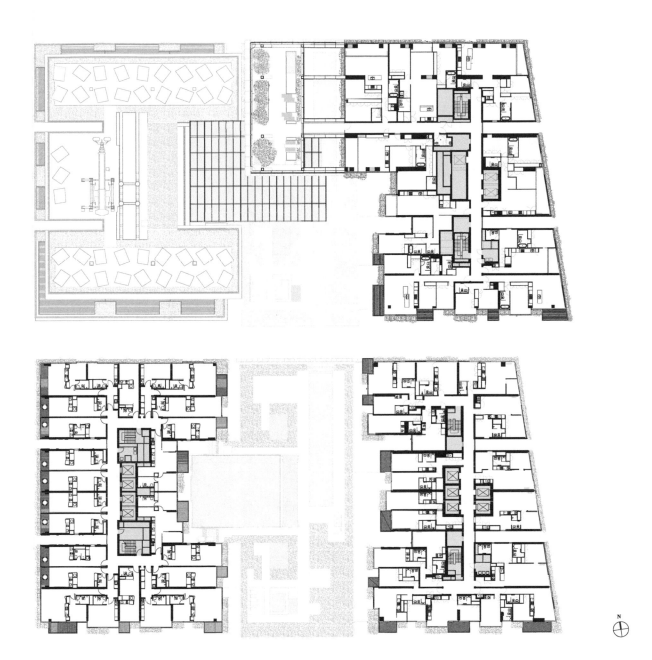

"There have been major advances in the incorporation of greenery in high-rise buildings over the past few years – but nothing on the scale of this building has been attempted or achieved. One Central Park points the way forward, not only for an essential naturalization of our built environment, but for a new aesthetic for our cities entirely appropriate to the environmental challenges of our age."

Antony Wood, Juror, CTBUH

Finalist
Best Tall Building Asia & Australasia

8 Chifley
Sydney, Australia

Completion Date: July 2013
Height: 141 m (461 ft)
Stories: 30
Area: 21,700 sq m (233,577 sq ft)
Use: Office
Owner/Developer: Mirvac Developments; Keppel Reit
Architect: Rogers Stirk Harbour + Partners; Lippmann Partnership
Structural Engineer: Arup
MEP Engineer: Arup
Main Contractor: Mirvac Constructions
Other Consultants: Arup (façade); Aspect Studios (landscape); Jenny Holzer (artist); Renzo Tonin & Associates (acoustics)

"The liberation of the ground plane through the choice of structural system for the tower is surprising and has a large positive impact on the public realm of the CBD of Sydney."

David Gianotten, Juror, OMA

The tower at 8 Chifley provides functional quality offices while creating opportunities for connectivity between occupiers from different parts of the building. The office spaces across 21 levels are connected by a series of adaptable two- and three-story interlinked vertical "villages." These villages, ranging in size from 1,800 to 2,600 square meters, provide the building with a high degree of flexibility, while creating a variety of individual workspace environments that allow privacy and interaction between individuals. This floor space, within a void and yet within the tower, allows the redistribution of space higher up the building where better views can be enjoyed.

The villages are interspersed with full-floor office levels, which allow for multiple villages to be connected. These dramatic vertical business units, each of up to four floors in volume, frame magnificent views over the cityscape. The larger "villages" create 45 percent more perimeter space with enhanced natural light, compared to traditional floor plates, providing flexibility for either open-plan or cellular-office layouts.

Central to the building's sense of connectivity and community is the building's social heart: its elevated "village square," on the 18th floor, set within a three-story void. This landscaped space contains a glass pavilion for all-weather use. The profiled roof also allows level 30 to take on similar qualities to the loggia spaces at ground level and at level 18. Framed by a wind-permeable structure that neutralizes the wind load, the light structure provides shading to the terrace, while allowing an open character.

On a prominent, north-facing site, 8 Chifley makes the most of its small site, which, unusually, is open on three sides. The new building reaches to the edges of its site and opens up at the lower six floors, forming a large public space that addresses one of Sydney's few existing city squares. This tight site presented a number of logistical construction challenges, which led to

"A superb office building that combines offices with meeting spaces in the sky, and public meeting spaces at the ground level. Colorful, joyous and meticulously detailed."

David Scott, Technical Jury Chair, Laing O'Rourke

extensive use of off-site manufacture and well-timed deliveries to keep to a tight program.

The design aims to create a building whose carbon emissions are at least 50 percent less than those of a "typical" Sydney CBD office, through a wide-ranging environmental design strategy. The blackwater recycling plant for the treatment and reuse of building and main water reduces the building's demand on potable water supply and discharges. A tri-generation system provides on-site base building power, heating and cooling, and peak load reduction from the existing electricity grid. The tri-generation system is also capable of exporting power to other buildings at certain times of the year.

A chilled-beam mechanical system reduces energy use and requires a high volume of fresh-air intake. The high-efficiency façade, including external shading and performance glazing reducing heat load, directs sunlight and daylight glare. A naturally ventilated ground floor glass lobby enclosure allows cooling without extensive energy expenditure. Basement facilities accommodate cyclists through bike racks, changing rooms and lockers, and the roof is designed to accommodate photovoltaic cells at a future date. In all, these strategies contribute to the building's 6 Star Green Star rating, the highest achievable in Australia.

Jury Statement

It is almost ironic that a building whose profile owes so much to the prominent display of musculature, in the form of brightly painted red steel cross braces and an exterior staircase that seems to bind the structure to an unseen back wall – can also provide such tranquil moments of repose. But then, this is what the best cities do, in all kinds of unplanned ways. The designers have consciously assembled the antipodal facets of city life, for a visceral experience in one building.

Finalist
Best Tall Building Asia & Australasia

Abeno Harukas
Osaka, Japan

Completion Date: March 2014
Height: 300 m (984 ft)
Stories: 60
Area: 212,000 sq m (2,284,949 sq ft)
Use: Hotel/Office/Retail
Owner/Developer: Kintetsu Corporation
Architect: Takenaka Corporation (design)
Structural Engineer: Takenaka Corporation (design)
MEP Engineer: Takenaka Corporation
Main Contractor: Takenaka Corporation; Okumura Corporation; Obayashi Corporation; Dai Nippon Construction; and The Zenitaka Corporation
Other Consultants: Bonbori Lighting Architect & Associates, Inc. (lighting); Hiromura Design Office (way finding); Pelli Clarke Pelli Architects (façade); Studio on Site (landscape); Takenaka Corporation (façade)

"This design formally realizes the bigger building as a city within the city, with all its diversity and complexity intact and visually expressed. Overall, a stunning project."

Sir Terry Farrell, Juror, Farrells

Abeno Harukas is the tallest building in Japan, but its significance extends beyond this, to its anchoring role in the urban core of one of the country's great cities, and for its novel use of greenery.

Abeno-Tennnoji railway station, which occupies the podium of the building, is a high-density hub where the number of passengers exceeds 70,000 a day. Abeno Harukas connects the metropolitan railway network to a high-density urban complex, incorporating a department store, art museum, school, hospital, office, hotel, observatory, and rooftop gardens. This multi-purpose network of services maximizes the performance of each function, and connects these programs with various vertical and horizontal circulation paths. In this compact and dense complex, the varied activities of 140,000 people energize not only this area, but also the metropolitan area along the railway network extending from the tower.

Previous Spread

Left: Overall view of tower in city

Right: View of void space and internal circulation

Current Spread

Right: Section highlighting voids, bracing, and circulation

Opposite Top: Sky garden on level 38 overlooking city

Opposite Below: View of the floor 60 observatory deck

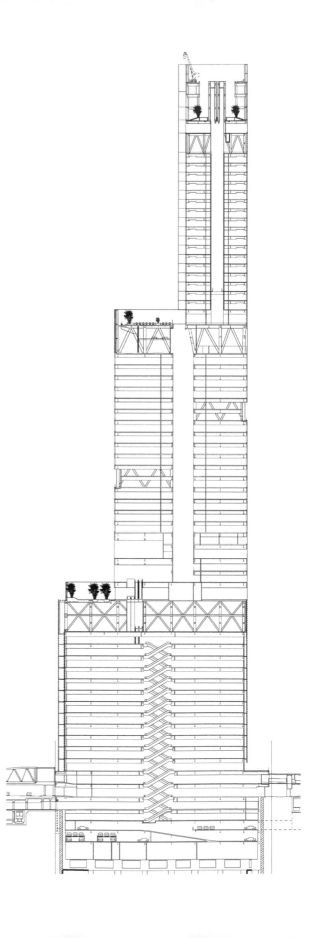

Sited in a high-density urban area, the shape of the large volumes comprising the tower were determined through various factors, such as impact of wind on the surrounding area, relation to the scale of the surrounding neighborhood, and circulation of occupants. The asymmetric structural megatruss, optimized to the program of the building, determines the void spaces, which offer space for vertical transportation as well as air circulation.

Three volumes with different floor areas are shifted and stacked, drawing sunlight and wind to the center void between offices, creating three-dimensional, cascading gardens. Further gardens placed on rooftop setbacks reconcile the vertical urban landscape with an adjacent park, while the semi-public gardens at the top of each volume are visible through the glass façade, forming a psychological connection to the ecology of the city. The diverse urban activities generated by the confluence of various functions inside transmit to the exterior through the transparent curtain wall. The scale of the tower is related to the existing micro-urban

"A well-considered structure with a seemingly effortless integration of its wide-ranging uses. Its volumes are integrated with a wide variety of indoor and outdoor spaces."

Jeanne Gang, Jury Chair, Studio Gang Architects

tissue through the use of public pedestrian paths on various floor levels.

Programmatic, structural, and environmental imperatives all intersect productively in this design. The truss frame installed on the upper levels, inspired by the central pillar design of traditional Japanese pagodas, also stabilizes the tower to withstand a 2,000-year earthquake. The voids inside the building are useful for ventilation and heat exchange. The department store's void channels waste heat inside ceilings and sends the cooled exhaust air to the upper floor's cooling tower by way of a buoyancy ventilation system. Voids in the office area intake natural light and wind to the central core section and render perimeter hallways as portico-like spaces. At night, cool fresh air is taken into a cool storage system, while hot air is purged.

Single-use buildings usually concentrate energy consumption during certain hours of the day. However, the multifunctional design of Abeno Harukas improves thermal efficiency and equalizes overall energy consumption, contributing to a significant reduction in CO_2 emissions. The building's multi-use design facilitates the incorporation of expansive energy-saving technologies. Waste heat generated throughout the year by air conditioning, essential to department store operations, is reused to produce hot water for the hotel above. Garbage from the restaurants and hotel facilities is effectively used for bio-gas power generation. As a result, CO_2 emissions will be reduced by 35 percent against comparable buildings.

Jury Statement

The elegance of Abeno Harukas is an achievement. Consider the seeming cacophony of functions packed into one building: hospital, art museum, retail, office, hotel, all atop a railway station that handles 70,000 passengers per day. Even more of a surprise is the amount of open, serene space in which one can catch a breath – the numerous atria, rooftop parks, and circulation spaces turn what could be a claustrophobic mash of programs into a vertical cross-section of city.

Finalist
Best Tall Building Asia & Australasia

Ardmore Residence

Singapore

Completion Date: September 2013
Height: 136 m (446 ft)
Stories: 36
Area: 17,178 sq m (168,627 sq ft)
Use: Residential
Owner/Developer: Pontiac Land Group
Architect: UN Studio (design); Architects 61 (architect of record)
Structural Engineer: Web Structures Singapore
MEP Engineer: J. Roger Preston Group
Main Contractor: Shimizu Corporation
Other Consultants: Arup (façade); Terry Hunziker (interiors)

"There are few high-rise buildings that blur the boundary between 'in' and 'out' at height like Ardmore does. The fact that large parts of it are also pre-fabricated is amazing."

Antony Wood, Juror, CTBUH

The Ardmore Residence in Singapore is located in a prime location close to the Orchard Road luxury shopping district and enjoys expansive views of the panoramic cityscape of Singapore and the green areas of its immediate surroundings. The primary concept for the design of the residential tower is a multi-layered architectural response to the natural landscape inherent to the "Garden City" of Singapore.

The façade is derived from micro-design features, which interweave structural elements, such as bay windows and balconies, into one continuous line. The façade pattern is repeated every four stories of the building, while rounded glass accentuates the column-free corners. Intertwining lines and surfaces wrap the apartments, seamlessly incorporating sun screening, while ensuring that the inner qualities of the apartments and the outer appearance together form a unified whole. From a distance, the tower appears to adopt divergent contours, whereas up close, a sense of organic mutation and transition is achieved as one circumnavigates the building.

The apartments embody the idea of a "living landscape." An indoor-outdoor living experience is established through the inclusion of large windows and double-height balconies. These elements offer views across Singapore, while the vertical balcony voids offset the horizontality of the more private interior spaces. With the terrace spaces integrated into everyday internal living scenarios, the links between interior and exterior spaces are made seamless. Bay windows create natural shading on the glass to minimize heat gain and provide opportunities for planting by the residents.

The floor plan of the apartments is designed to increase the amount of daylight and take advantage of the panoramic views. The plan is based on an analysis of Renaissance villa prototypes and the concept of a "cours d'honneur" as in-between spaces that enable visual links between different parts of the same house. Here, this concept is appropriated and translated in

"The design of the Ardmore Residences is innovative while respecting the principles of efficiency and sustainability. The outcome is a special project in the 'towerscape' of Singapore."

David Gianotten, Juror, OMA

Previous Spread

Left: Overall view of tower

Right: Interior double height open air terrace

Current Spread

Below: Typical residential space

Opposite Top: Typical floor plan

Opposite Bottom: View of tower from base

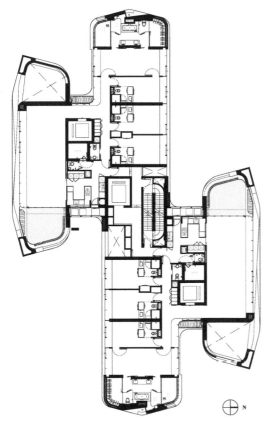

order to allow for visual interconnection between specific areas within the apartments. This concept furthermore increases privacy for sleeping rooms and adds a four-dimensional aspect to the layout, as both wings can operate separately, service different needs and be active at different times of the day.

An open framework is introduced at the base of the tower, which enables full connectivity and transparency across the ground level landscaping, while organizing the shared amenity facilities. While building regulations in Singapore specify both the height and area of high-rise buildings, views are also deemed to be essential for the occupants. The raised design of the Ardmore Residence integrally incorporates these parameters to take advantage of the potential they afford to optimize the design concept and simultaneously create a fully integrated living and leisure landscape for its occupants.

The Ardmore Residence has been built using reinforced concrete, involving a substantial amount of prefabrication work. Single-story shear walls cantilever from the inner core walls and support one floor above and one floor below at the same time. An interlocking system was developed to stagger these cantilevered shear walls across the height of the tower to produce the building's distinctive look. Combined with the rounded corners of the walls and suspended ceilings, the building is imbued with the romance of an ocean liner, sailing over the urban greenery of Singapore.

Jury Statement

For all the intensity of its undulating, interlocking folds, coves, and grilles, the Ardmore Residence is all the more remarkable for what it is not – a bespoke work of sculpture constructed at great expense. Instead, it is a testament to what prefabrication can accomplish when the "kit of parts" is well conceived, yielding not only exterior visual interest, but also floor plans that liberate the living rooms from the core of the tower, offering residents maximum exposure and views.

Finalist
Best Tall Building Asia & Australasia

FKI Tower
Seoul, South Korea

Completion Date: December 2013
Height: 245 m (804 ft)
Stories: 50
Area: 116,037 sq m (1,249,012 sq ft)
Use: Office
Owner/Developer: Federation of Korean Industries
Architect: Adrian Smith + Gordon Gill Architecture (design);
Chang-Jo Architects (architect of record)
Structural Engineer: Thornton Tomasetti; DongYang Structural Engineers Co., Ltd.
MEP Engineer: Environmental Systems Design, Inc.; Hanil MEC. Engineering Co., Ltd
Main Contractor: Hyundai Engineering & Construction; STX JV
Other Consultants: Adrian Smith + Gordon Gill Architecture (sustainability);
Construction Cost Systems (cost); Fortune Consultants, Ltd. (vertical
transportation); Lerch Bates (façade maintenance); Rolf Jensen & Associates
(fire); RWDI (wind); Shen Milsom Wilke, Inc. (acoustics); SWA Group (landscape);
Thornton Tomasetti (façade); V3 Companies (civil)

"This building repeats a logical solar strategy across its façades. At the same time it achieves a remarkable aesthetic, all the more powerful because it is driven by performance."

Jeanne Gang, Jury Chair, Studio Gang Architects

The new head office for the Federation of Korean Industries (FKI) is a major new addition to the skyline of Seoul, Korea. The tower features an innovative exterior wall, designed specifically for the project. This unique skin helps reduce internal heating and cooling loads and collects energy through photovoltaic panels that are integrated into the spandrel areas of the southeast and southwest faces. FKI clearly illustrates the advancement in building façades from simple wall systems to high-performance, integrated architectural and engineering design solutions.

FKI's unique exterior wall system combines maximum access to views, energy efficient strategies and energy generation technologies. By angling the spandrel panels 30 degrees toward the sun, the amount of energy collected by the photovoltaic panels is maximized. Below the spandrel panels, the vision panels are angled 15 degrees toward the ground, minimizing the amount of direct sun radiation and glare to the interior space.

Previous Spread

Left: Overall view of tower with adjacent green space

Right: Rooftop garden showing PV panels

Current Spread

Below: View of the sculptural podium containing public use amenities

Opposite Top Right: Section depicting strengths of folded exterior façade

Opposite Bottom Right: View of one of the multi-story sky garden atria distributed throughout the building's height

The use of building integrated photovoltaic panels (BIPV) was seen as an architecturally appealing way to meet a strict zoning requirement that 5 percent of the building's energy be created on-site, while the optimization of the panels became a driving factor in developing the architectural expression. The local electric utility company (KEPCO) provided a favorable 5-to-1 buy-back rate for on-site green-energy generation. The payback for the BIPV panels, which would have typically been 30–35 years, was reduced to about seven years, due to these incentives.

As part of the initial design process, an Ecotech model was used to determine the optimal areas for BIPV on an orthogonal building, given the surrounding buildings which partially shade the site. It was determined that BIPV would be best used on the southeast and southwest faces, above Level 14. Not all sides of the building are currently suitable for BIPV; therefore, it was important to design panels that were interchangeable with insulated spandrel panels for maximum flexibility, without a significant change in the symmetry of building

Jury Statement

While many tall buildings trade on their iconicity from a distance, FKI Tower is a building that rewards the visitor more on closer inspection. Both the bulbous glass podium and the tilted spandrel panels of the main tower invite the outside viewer in, as much as they let verdant views and indirect light inside. The folded exterior plays a vital role in bringing electric energy to the building through solar panels, while energizing the façade visually, regardless of the side of the glass on which one stands.

> *"It's a relief to find a building that is so formally simple and which avoids shape making as an end in itself. The cladding is the star – clever and inventive."*
>
> Sir Terry Farrell, Juror, Farrells

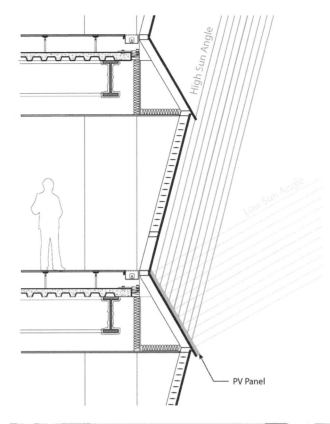

PV Panel

expression. The design also allows for the panels to be replaced, as new technology that can take advantage of indirect light becomes available.

The result is a unique folded exterior texture that is both purposeful and visually distinctive. Benefits include a reduction of glare and heat gains from direct sunlight, while maintaining a high level of indirect light. Through most of the day, the building is able to use the geometry of the exterior wall to self shade the perimeter spaces that would normally be inundated with direct sunlight.

FKI features several multi-story garden atria throughout the height of the building, as well as an expansive rooftop atrium garden that also has custom photovoltaic panels. As with the exterior wall panels, the ideal angle of the panel placement on the roof was studied in detail. Within the limited area of FKI's roof, it was determined that a 10-degree angle allowed for more panels to be installed closer together, minimizing the effect of the panels casting shadows on each other and ultimately producing more solar energy for the building, while still allowing enough light through for the planting below.

The sculptural podium's amenities, available for public use, include a banquet hall, a central restaurant and a conference center. Site orientation was altered during the course of design, sliding the tower to the west. This opened up more site frontage and allowed the podium to be closer to the street, giving it a presence it would not have had if located behind the tower, and allows more space for outdoor public landscaped courtyards.

Finalist
Best Tall Building Asia & Australasia

IDEO Morph 38
Bangkok, Thailand

"The simple rectangular shape is enhanced by the soft landscape applied to the exterior. The balcony projections enrich the texture of the façade, all creating a harmony with the nearby garden."

Wai Ming Tsang, Juror, Ping An Development

The Ideo Morph 38 development is located away from the high density and congestion of Sukhumvit Road, in a blissfully green low-rise residential area. Its ordered, pixellated façades provide a contrast to the visual clutter characteristic of central Bangkok. The development is separated into two towers to maximize the building plot ratio, with each building targeted to different demographics.

The lower tower of the two, Skyle, contains duplex units targeted to singles or young couples, offering the smaller footprint of the two. These duplex units are expressed vertically to achieve a generous 5.4-meter floor-to-floor height. Due to limitations on unit size and variety, the balconies and the air-condensing units on the outside of the building contribute to create a "pop-up" effect through variation in the façade.

In contrast the taller tower, Ashton, emphasizes cantilevered spaces, with units targeted at families. The unit sizes and types vary from a single bed

Completion Date: January 2013
Height: Ashton: 132 m (433 ft); Skyle: 64 m (210 ft)
Stories: Ashton: 35; Skyle: 12
Use: Residential
Owner/Developer: Ananda Development PCL
Architect: Somdoon Architects
Structural Engineer: H. Engineer Co., Ltd (design); Westcon Co., Ltd (engineer of record)
MEP Engineer: Elemac Company limited (design); Mect Co., Ltd. (engineer of record)
Project Manager: MJR Management
Main Contractor: Westcon Co., Ltd
Other Consultants: Dot Line Plane (interiors); Flix Design (interiors); Meinhardt (façade); Shma Company Limited (landscape); Thai Thai Engineering (environmental)

Previous Spread

Left: Overall view of the towers

Right: View looking up between the towers highlighting the green façades

Current Spread

Left: Cantilevered tree balconies up the façade of Skyle Tower

Opposite Top: Upper and lower floor plan of the level 31 loft in Ashton Tower

Opposite Bottom: Interior view of duplex unit in Skyle Tower

"An unusually rich range of interior and exterior spaces is achieved. Living walls, patios and gardens are integrated vertically, providing great indoor-outdoor connections."

Jeanne Gang, Jury Chair, Studio Gang Architects

with a reading room, to duplex units with a private swimming pool and garden on the eighth floor, and a four-bed duplex penthouse at the top level. A 2.4-meter cantilevered living space projects from each of the units on the north side, made up of a glazed enclosure on three sides, providing maximum, unobstructed views. Each unit on the south has a semi-outdoor balcony, which acts as flexible space. The double layer of sliding windows allows for a transition between a conventional balcony and an extended, indoor living area.

Jury Statement

Through pushed and pulled volumes, planted balconies and interpolating skins of vertical greenery and solid panels, Ideo Morph 38 provides a transition from a low-rise, leafy neighborhood to a district of traditional high-rises. Access to greenery and views are not mutually exclusive. The combination of shelter and prospect is not the exclusive reserve of suburban single-family homes. It is not necessary to dwell in a large space to get a feeling of expansiveness.

The two towers are visually interconnected through a folding "Tree Bark" envelope that wraps around both towers. This outer skin is a combination of precast concrete panels, expanded mesh, and planters. The functions of the skin vary from acting as sun-shading devices to covering air-condensing units, while the "bark" elements on the west and east side strategically support green walls in accordance with the tropical sun's orientation. The green walls provide the residences and neighboring buildings with a comfortable visual and natural environment. The buildings are oriented in the east-west direction, helping to reduce solar heat gain.

The sky gardens found on both buildings at regular intervals also contribute to this design language, where the development becomes a "vertical landscape" and a part of the natural green surroundings. The addition of significant planter-based trees on cantilevered balconies adds the option of standing at height outdoors under a canopy, and some variation to what otherwise could be a monolithic exterior appearance. This also produces

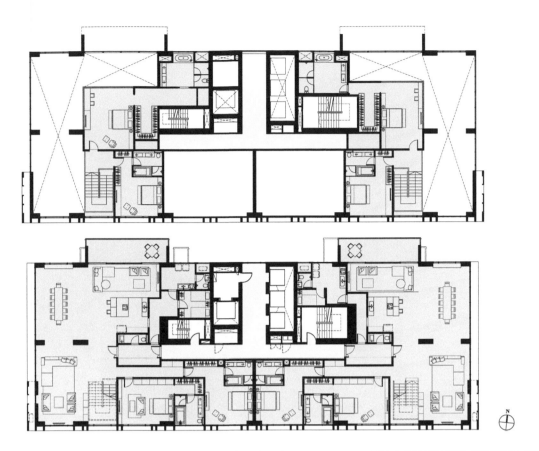

a special sensation on certain of the balconies: that of standing out apart from the building, yet enclosed beneath the shelter of leaves. Additionally, the parking garage in the podium is shielded by green walls.

The green walls are fed by planter boxes located behind expansion mesh, and are installed approximately 625 millimeters from the building envelope, permitting natural ventilation and providing a service corridor for maintenance. A vine system was selected for the green wall, due primarily to the fact that it is easily maintained and cost effective. The vine vegetation is also able to withstand the strong winds and robust weather conditions of Bangkok. The height of each floor varies from three to six meters, which provides sufficient height for the plants to climb and merge into one continuous, green surface.

Finalist
Best Tall Building Asia & Australasia

Sheraton Huzhou Hot Spring Resort

Huzhou, China

"The project manages to be both futuristic in its expression, and evocative of historic Chinese culture in its lake reflection of the number eight."

Antony Wood, Juror, CTBUH

Completion Date: December 2012
Height: 102 m (335 ft)
Stories: 24
Area: 30,799 sq m (331,518 sq ft)
Use: Hotel
Owner: Sheraton Huzhou Hot Spring Resort
Developer: Feizhou Group
Architect: MAD Architects
Structural Engineer: China Majesty Steel Structural Design Co., Ltd.
MEP Engineer: China Majesty Steel Structural Design Co., Ltd.
Main Contractor: Shanghai Xian Dai Architecture Design (Group) Co., Ltd.
Other Consultants: EDSA (landscape); Zhejiang Zhongnan Curtain Wall Co., Ltd. (façade)

The Sheraton Huzhou Hot Spring Resort hotel is located next to Nan Tai Lake in Huzhou, a city situated west of Shanghai and north of Hangzhou, with views to Suzhou and Wuxi across the lake. Since ancient times, Huzhou has been known as "the house of silk" and "the land of plenty," and is the only ancient cultural city in the Nan Tai Lake region. The hotel connects with this local context, referencing both traditional and modern conditions, and is iconic in its integration and reflective dialogue with the waterscape of Tai Lake.

The building takes full advantage of its waterfront by directly integrating architecture and nature. The circular building corresponds with its reflection in the water, creating a surreal image and changing connection between solid reality and an aquatic phantom. Beneath the sunlight and the reflection of the lake, the curved shape of the building is crystal clear. When night falls, the entire building is brightly illuminated by both its interior and exterior

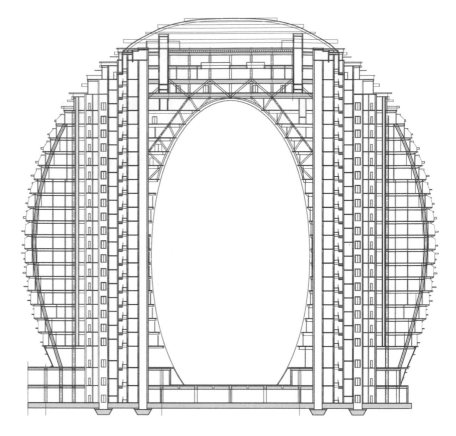

"The elliptical form crafts a simplistic but unique sculpture, while the structural system remains very simple and minimalistic."

Wai Ming Tsang, Juror, Ping An Development

lighting. Soft light wraps around the hotel and the water, resembling the bright moon rising above the lake, blending classical China and modern China in an interlocking embrace.

From the beginning, designing an inhabitable vertical glass ring posed a great challenge for the structural engineers. It was eventually decided that a reinforced concrete-core tube would be most lightweight, most capable of containing a dense program, and most earthquake-resistant. However, it was also a challenge to implement a concrete structure and meet the goal of reducing environmental pollution during construction. The mesh-covered, curved-surface structure gives the building its necessary rigidity, which is further enhanced by the bridge-like bracing steel structure that connects with the double-cone structure at the top floor. The hotel façade is covered with layers of fine-textured white aluminum rings and glass, creating a sense of drama and ambiguity around the building's scale.

The thin shape of the structure allows for good ventilation to the various rooms and suites. Each floor also has open-air terraces, shielding the interior from direct sunlight during the summer, while allowing it to penetrate the interior during winter. It is common to see people enjoying the sunlit warmth from their balconies during the summer. The outdoor space at the top of the structure acts to supplant the footprint at the ground by providing a roof garden for the enjoyment of guests.

The ring-shaped hotel provides all rooms with favorable views of the waterfront and surrounding city, while allowing for maximal natural light in all directions. The arc-like public space at the top of the building has great open views and acts as a "mid-air" multi-purpose room for large-scale activities. The experience of being at the top invokes a feeling of floating over the lake.

Through all of these gestures, the building emphasizes the harmony between man and nature, and enhances visitors' sensual and spiritual experiences.

Jury Statement

With its smooth curves and improbable proportions, this surreal concrete-tube structure evokes endless contemplation. Seen against the water, it does seem to be an expression of continuity with something below the surface that beguiles the tower's standoff with gravity and speaks to invisible forces. It is rare that a tall building can be both an artifact of human ingenuity and seem eternal and naturalistic, but the Sheraton Huzhou Hot Spring Resort accomplishes this.

Finalist
Best Tall Building Asia & Australasia

The Jockey Club Innovation Tower

Hong Kong, China

Completion Date: August 2013
Height: 71 m (234 ft)
Stories: 15
Area: 28,000 sq m (301,389 sq ft)
Use: Education
Owner/Developer: The Hong Kong Polytechnic University
Architect: Zaha Hadid Architects (design); AD+RG Architecture Design and Research Group (architect of record); AGC Design (architect of record)
Structural Engineer: Arup
MEP Engineer: Arup
Main Contractor: Shui On Construction & Materials
Other Consultants: Arup (façade, fire, geotechnical); Ho Wang & Partners Ltd. (traffic); Rider Levett Bucknall (quantity surveyor); Team 73 Hong Kong Ltd (landscape); Westwood Hong & Associates Ltd (acoustics)

"This is a gem of a building in Hong Kong, and although relatively small at 15 stories when compared to other towers in the city, it takes a dynamic and even imposing stance."

David Scott, Technical Jury Chair, Laing O'Rourke

The Jockey Club Innovation Tower is a new school of design building for the Hong Kong Polytechnic University which offers a creative and multidisciplinary environment. The building is located on a very tight and irregular site on the north side of the campus. It creates an accessible urban space which transforms how Hong Kong Polytechnic University is perceived and the way it uses its campus. The building projects a vision of possibilities for its future, as well as reflecting on the history of the university by encapsulating in its architecture the process of change.

The project re-examines and addresses a creative, multidisciplinary environment by collecting together the variety of programs of the School of Design. Having undergone a strict process of examination of the multiple relationships among its unique identities, these programs are arranged in the tower in accordance with their "collateral flexibilities."

"Marking the next step for design in Hong Kong, the building has changed the center of gravity on the university's campus and the image of the entry point of the central cross harbor tunnel."

David Gianotten, Juror, OMA

The fluid character of the Jockey Club Innovation Tower is generated through an intrinsic composition of its landscape, floor plates, and louvers, dissolving the classic typology of the tower and podium into a seamless composition. Internal and external courtyards create new spaces of an intimate scale, which complement the large open exhibition forums and outdoor recreational facilities to promote a diversity of civic spaces, integrated with the university campus.

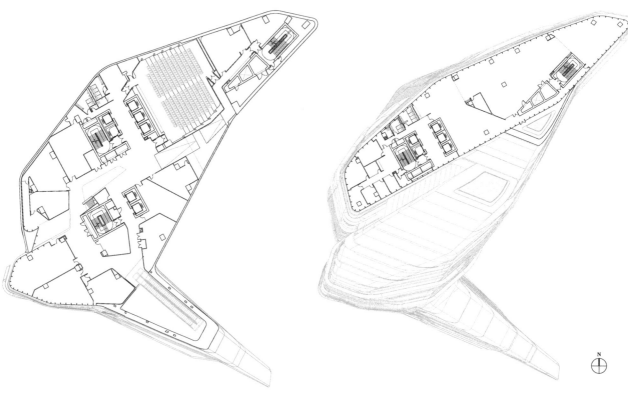

Inside, the building is configured as a place for learning, exchange, and synergy; it is at once flexible, open, and transparent for its staff and students. A series of fixed sun-shading louvers protects the building from excessive solar gain, while allowing maximized indirect natural daylight into its workplace. A series of maintenance walkways are implemented behind these louvers, with access from the building's interior. The tower cantilevers over a footpath north of the site, and this path could not allow for any foundations, thus the superstructure was creatively engineered by using the three main cores and beam-column frames for lateral load and eccentric tower loads. The concrete superstructure adopts a strategy of raking walls and columns, with discrete transfer beams to free the lower public levels from structural obstructions.

The new pedestrian level for the tower has been created as an open public foyer that channels deep into the building. The integrated pathway from Suen Chi Sun Memorial Square guides visitors to the main entrance. From here, a welcoming public space provides access to shops, cafeteria, and a museum through a generous series of open exhibition and showcase spaces, which span between the campus podium level and the ground floor. From the entry foyer, staff, students, and visitors move upwards through the various levels of openly glazed studios and workshops. The many studios and workspaces accommodated within the new School of Design appear as a variety of visual showcases. The route through the building becomes a transparent cascade of exhibition and event spaces – allowing the student or visitor to visually connect and engage with the work and exhibits. These routes promote new opportunities of interaction between the diverse user groups. In this way, the programs of the tower, comprising learning clusters and central facilities, generate a dialogue between respective spatial volumes and disciplines of design.

The energy and life of the school is reformed vertically, embodying an environment which can nurture design talents in a collective research culture, where many contributions and innovations can influence each other as a community.

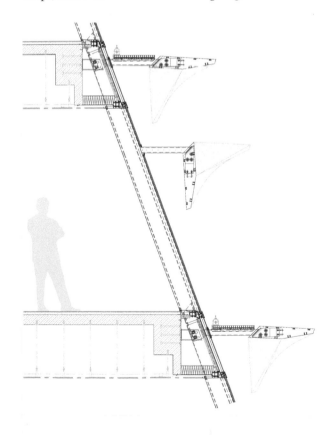

Jury Statement

This building presents a sculptural, organic and iconic image to its busy city context. Despite its irregular shape, the plan enables a high level of functionality. Plans are laid out according to the functional requirements, allowing effective organization, while the circulation system and spatial arrangement of the interior are straightforward. Interior and exterior architectural elements help the building break away from the relentless urban grid and the "textbook" institutional campus typology.

Previous Spread

Left: Overall view of tower from the northeast

Right: Interior open circulation space

Current Spread

Opposite Top Left: Overall view from southeast

Opposite Bottom: Floor plans – level 3 (left) and level 13 (right)

Left: Façade detail section

Wangjing SOHO
Beijing, China

Completion Date: September 2014
Height: Tower 1: 118 m (387 ft); Tower 2: 127 m (417 ft); Tower 3: 200 m (656 ft)
Stories: Tower 1: 25; Tower 2: 26; Tower 3: 45
Area: 392,265 sq m (4,197,235 sq ft)
Use: Office
Owner/Developer: SOHO China Co. Ltd
Architect: Zaha Hadid Architects (design); China Construction Design International (architect of record)
Structural Engineer: CABR (design); China Construction Design International (engineer of record)
MEP Engineer: Arup (design); China Construction Design International (engineer of record)
Main Contractor: China State Construction Engineering Corporation
Other Consultants: Arup (façade); Ecoland (landscape); EMSI (LEED); Ikonik (way finding); Inhabit Group (façade); Lightdesign (lighting); Yonsei University (wind)

> *"Wangjing SOHO creates one of those stop-dead moments for the Beijing visitor – sculptural mountains in the middle of the city."*
>
> Antony Wood, Juror, CTBUH

The Wangjing SOHO Project is designed as three dynamic mountain- or fish-like forms, pulling flow through the site with their convex forms. The juxtaposition of the towers affords a continuously changing, elegant, and fluid view from all directions. The exterior skin of the towers consists of flowing, shimmering ribbons of aluminum and glass that continuously wrap around the buildings and embrace the sky, threading through a landscape with approximately 60,000 square meters of green area open to the public. Inspired by the surrounding movement of the city, and the sun and wind, the project aims to lend a strong identity to the Wangjing area, creating a gateway-beacon that can be seen by travelers along the highway heading to or from Beijing Capital International Airport.

The site for the Wangjing SOHO Project is located in the Chaoyang District of northeast Beijing, between Fourth and Fifth Ring Roads. The area contains the offices of many Chinese startup companies, as well as global companies such as Microsoft, Daimler,

Caterpillar, Panasonic, Nortel, and Siemens. It is conveniently located on the way to the airport and near various metro stations, and is home to a vibrant mix of local and international residents and visitors.

The building program contains offices and retail above grade, retail below grade in B1 basement level, and parking and mechanical in the B2, B3 and B4 basements. The composition of the towers extends into the surrounding landscape, with flowing lines creating paths of movement and exciting activity zones of shopping and leisure. The lines of movement extend to the perimeter and integrate all the green areas around the site. Between the main building towers is a "canyon" of retail shops and activities, and several pavilion gate buildings that create a shopping street at the ground level. There are two sunken garden courts east and west of the canyon that continue the landscaped paths down to the retail concourse below.

"The soft fluid forms of the buildings are reminiscent of the undulating geological structures found in nature, and the relationship between the towers changes with every perspective."

David Scott, Technical Jury Chair, Laing O'Rourke

The main tower entrance lobbies, facing outwards to the city, welcome visitors into large dynamic halls that direct one into the office tower floors above, and to the breezeway and retail levels at the second floor and sunken garden levels below. Up above in the office towers, there are simple open-plan office spaces offering natural daylight and continuous panoramic views in all directions. Most of the roofs are covered with louvers and the top of the roof surfaces are coated with highly reflective material, in order to mitigate the heat-island effect in the city. The buildings have double-insulated unitized glazing systems and horizontal bands of white aluminum that provide overhangs for sun-shading, while providing maintenance terraces and water collection.

To encourage more sustainable transportation access, special parking spots are reserved for low-emission cars. Bicycle parking and shower facilities are also provided. Direct access to subway stations and bus stops nearby have been integrated into the planning.

For better indoor environment quality for the occupants, the fresh air rate per person provided exceeds the ASHRAE standard by 30 percent. Highly efficient filters are installed to remove PM2.5 particles in the AC system. In the interior design, low-volatile organic compound (VOC) materials are carefully chosen to eliminate pollution from the outset.

Previous Spread

Left: Overall view of towers in urban context

Right: Entrance lobby with lighting detail

Current Spread

Opposite Left: Building Section

Opposite Right: Overall view from East

Right: Typical floor plans

Jury Statement

Wangjing SOHO presents a welcome, refreshing essay in sinuousness, biomorphism and nuanced open space. Its tapering forms and towering lobbies at once recall the romance and optimism of Jet-Age architecture, while speaking to a future in which the optimum solutions for enclosing mankind's endeavors will perhaps not be rectilinear "machines for working," but will instead be "conversations with the primordial."

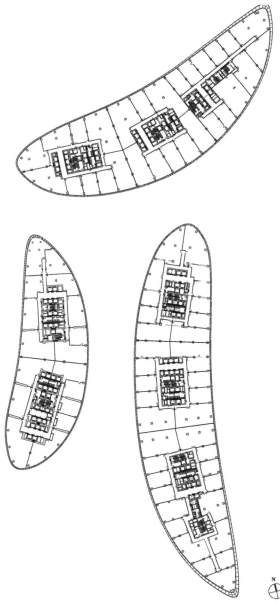

41X
Melbourne, Australia

The 41X project was developed by the Australian Institute of Architects to create new space for its long-term use in the center of Melbourne. The building accommodates a dense program on a small site area, with rental offices above the headquarters and public display area for the AIA. The small floor plate enables small businesses to create their own identity within the building. As opposed to the common trend of moving to low-rise, large floor-plate buildings, this development offers boutique-sized floor plates of 280 square meters each, on a site area of 330 square meters.

As a typical tower/podium setback arrangement was impossible, the design "chisels" the lower levels to create intimate public spaces and conceptually draw the city street up into the building through a feature stair at the perimeter. The building's "Sustainability Charter" will be incorporated into the owner's corporation rules and mandates the use of 100 percent green power and minimum commitments to carbon offsetting through the Australian Carbon Trust.

Completion Date: January 2014
Height: 85 m (279 ft)
Stories: 22
Area: 7,000 sq m (75,347 sq ft)
Use: Office
Owner/Developer: Australian Institute of Architects
Architect: Lyons
Structural Engineer: Winward Structures
MEP Engineer: AECOM
Project Manager: DPPS Projects
Main Contractor: Hickory Group
Other Consultants: Aurecon (fire); AECOM / Davis Langdon (quantity surveyor)

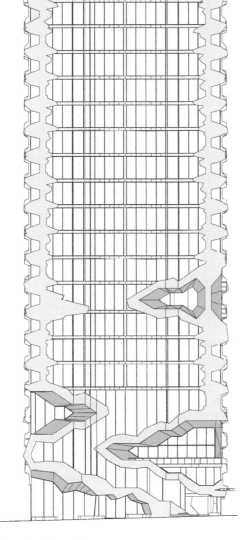

Opposite: Overall view of tower from northeast

Above: South elevation

Top Right: Effect of façade from street

Bottom Right: Balcony and feature stair from street corner

171 Collins Street
Melbourne, Australia

Developed directly out of the richness of its inner city location, 171 Collins Street is a Premium Grade 6 Star Green Star office building. The crystalline tower provides an iconic focal point to Melbourne's city skyline, featuring an impressive nine-story internal glass atrium, premium finishes and facilities, efficient side core floors, a woven crystalline façade and the latest in technology and environmental design.

Its design has four main components: the Mayfair Building, the Flinders Lane podium, the atrium, and the office tower. Together, these form a unified whole. The Collins streetscape is maintained by setting the tower form back 32 meters from the street, while the Flinders Lane façades emulate the textures and scaling devices of the surrounding brick warehouse buildings. The new 171 Collins Street complex lends an elegant but not overpowering backdrop to the historic Mayfair Building, whose façade continues to enclose the Collins Street entrance to the project.

Completion Date: May 2013
Height: 86 m (282 ft)
Stories: 18
Area: 33,132 sq m (356,630 sq ft)
Use: Office
Owner/Developer: Charter Hall; Cbus Property
Architect: Bates Smart
Structural Engineer: Winward Structures
Project Manager: APP Corporation
Main Contractor: Brookfield Multiplex
Other Consultants: Bates Smart (interior); BG&E Façades (façade); Lovell Chen (preservation); Rider Levett Bucknall (quantity surveyor); Umow Lai (environmental); Urbis (planning); Winward Structures (façade)

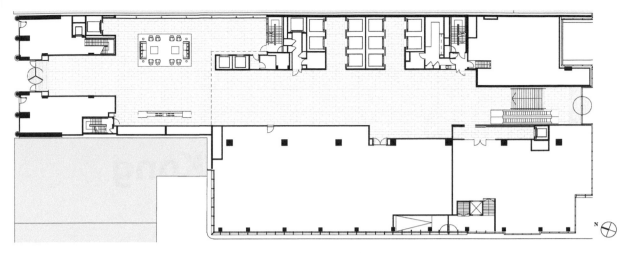

Opposite: View of preserved historic Mayfair building façade with new tower beyond

Top: Ground floor plan

Left: Overall view of tower from south

Right: Lobby atrium

Academic 3
City University of Hong Kong
Hong Kong, China

Academic 3 connects sectors of the City University of Hong Kong campus and offers a green oasis at its heart. Facilities include classrooms, teaching and research laboratories, multifunctional rooms, common areas, administrative offices, a 600-seat lecture theater, a canteen, and a roof garden. In the podium, teaching labs predominate the program, while in the upper section, the administration building contains the board council and Vice Chancellor's offices, as well as two sky courts, which afford prime views.

The green deck connects the dormitory in the northern campus to the university core, and grows from two to four stories as it progresses toward Cornwall Street. Beneath the open park at roof level, the podium, known as the "Forest of Intellect," which provides informal meeting space and amenities, accommodates teaching rooms and the lecture theater at its southern tip. By raising the podium above ground level, pedestrian access from the street to the park beyond also gives the public a greater natural resource to enjoy.

Completion Date: March 2013
Height: 107 m (351 ft)
Stories: 19
Area: 42,100 sq m (453,161 sq ft)
Use: Education
Owner/Developer: City University of Hong Kong
Architect: Ronald Lu & Partners
Structural Engineer: Hyder Consulting
Main Contractor: Hsin Chong Construction Company Limited
Other Consultants: ACLA (landscape); Langdon & Seah (quantity surveyor)

Opposite: Overall view of tower from southwest

Top: Green deck stretches beyond the tower to the north

Left: Interior circulation

Right: View of the tower from the green deck

Nominee
Best Tall Building Asia & Australasia

Albert Tower
Melbourne, Australia

A strong sculptural form with a unique identity, Albert Tower's design is inspired by its proximity to Melbourne's Botanic Gardens. The building is cloaked in a hexagonal exoskeleton, the language of which references the cellular structure of plants. When seen from a distance the clarity of the exoskeleton generates a singular expression, reinforcing a simple understanding of the complete form. This apparent simplicity lends the building a strong identity and appropriate scale in the broader city context. On closer approach, the interplay of the balconies and exoskeleton encourages further inspection.

An extension of the exoskeleton to the ground anchors the building to its context. On the top floor, comprehensive communal facilities are designed to foster social interaction between residents. A consistent and detailed approach to the interior design language is found throughout the common areas and apartments. This is further enhanced by the legibility of the tower's façade inside the apartments, strengthening the connection and identity felt by each resident to the building.

Completion Date: October 2013
Height: 93 m (305 ft)
Stories: 30
Area: 21,175 sq m (227,926 sq ft)
Use: Residential
Owner: Perri Projects; The Carter Family
Developer: Perri Projects
Architect: ROTHELOWMAN
Structural Engineer: Winward Structures
MEP Engineer: Wood and Grieve Engineers
Project Manager: Gallagher Jeffs
Main Contractor: Icon Construction Australia
Other Consultants: Acoustic Logic (acoustics); MEL Consultants Pty., Ltd. (wind); Reddo (quantity surveyor); Slattery Australia (cost); Thomas Nicolas (fire); Wood and Grieve Engineers (environmental)

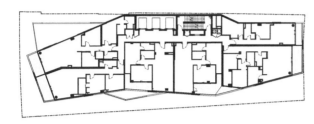

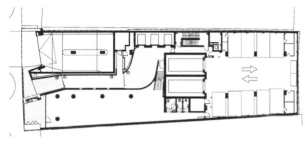

Opposite: Overall view of tower from northeast

Top Left: Floor plans – typical residential floor (top) and ground floor (bottom)

Bottom Left: Typical apartment space

Top Right: Tower entrance

Bottom Right: Façade detail

Nominee
Best Tall Building Asia & Australasia

Anhui New Broadcasting & TV Center
Hefei, China

Standing at the south end of Swan Lake in Hefei, Anhui New Broadcasting & TV Center is the signature headquarters of the Anhui region's largest television producer, while offering the highest observation point in the region. Its elongated, curved form optimizes circulation between functional and open areas by providing various lobbies and public spaces for multifunctional use. The tower's facilities are divided into two different sections: the lower portion is home to all broadcasting studios, multimedia companies and technical service areas, while the upper portion contains offices. The building features the largest multifunctional broadcasting hall in Asia, at 3,600 square meters.

The innovative structure showcases fluidity through its swift upward momentum, recalling an unfolding paper scroll. This spiral shape conceptualizes ancient Chinese wisdom as physical representations of local Li style fonts; traditional Chinese calligraphy adorn the outer skin of its double-layered façade. The outward culturally traditional influences offer a unique contrast to the modern, high-tech functions of the Centre itself.

Completion Date: May 2013
Height: 245 m (778 ft)
Stories: 46
Area: 190,582 sq m (2,051,408 sq ft)
Use: Office
Owner/Developer: Anhui Broadcasting & TV Station
Architect: NewDesignArchitecture (design); China Academy of Building Research (design)
Structural Engineer: China Academy of Building Research
MEP Engineer: China Academy of Building Research
Main Contractor: The Second Construction Co., Ltd. of China Construction Third Engineering Bureau
Other Consultants: Beijing Special Engineering Design and Research Institute; National Center of Performing Center (acoustics); Wuhan Lingyun (façade)

Opposite: Overall view of tower

Top: Overall view of entire project showing extensive podium "tail"

Above: View of lobby in middle of tower form

Bottom Right: Overall view of tower and beginning of curved form

Baku Flame Towers
Baku, Azerbaijan

The Baku Flame Towers have become a global symbol for Baku and Azerbaijan. The flame-shaped silhouette of the buildings was selected to pay homage to Azerbaijan's nickname as the "land of fire," a result of the area's rich deposits of natural gas — which continue to drive the region's economic development.

The design consists of three towers, each containing a distinct function. The tallest tower contains 130 residential units, a 250-room hotel occupies the second, and commercial office space takes up the third. The podium contains additional retail space on three levels, located directly above the hotel's ballroom and conference facilities. Two of the three levels of retail are accessible from the exterior, creating a unique cascade of space across the sloped site. The buildings present the opportunity for visitors to move through the shopping mall and walk from one tower to the other, change levels, or walk across the top of the landscaped podium.

Completion Date: June 2013
Height: Tower 1: 182 m (596 ft); Tower 2: 165 m (540 ft); Tower 3: 161 m (528 ft)
Stories: Tower 1: 39; Tower 2: 36; Tower 3: 28
Total Area: 249,070 sq m (2,680,967 sq ft)
Use: Tower 1: Residential; Tower 2: Hotel; Tower 3: Office
Owner/Developer: Azinko Development MMC; DIA Holdings
Architect: HOK
Structural Engineer: Balkar Mühendislik
MEP Engineer: Adana Dinamik Mühendislik; HB Teknik
Project Manager: Hill International
Main Contractor: DIA Holdings
Other Consultants: Frances Krahe & Associates, Inc. (lighting); Gill (landscape)

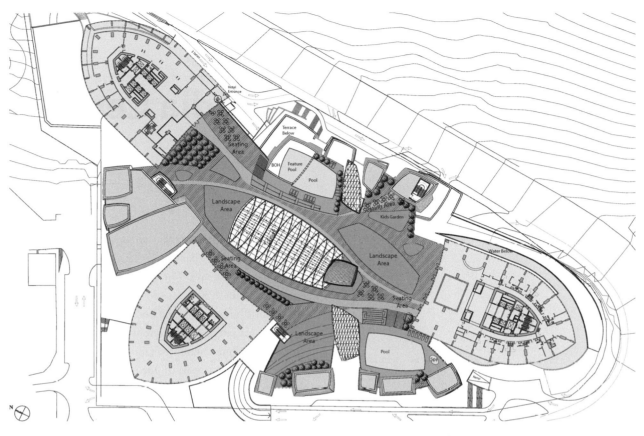

Opposite Top: Overall view of towers, daytime

Opposite Bottom: Overall view of towers, illuminated at night

Top: Podium floor plan

Above: Podium space with water feature

Right: Plaza on podium and relation to tower base

Changzhou Modern Media Center
Changzhou, China

Changzhou Modern Media Center is located in the Xinbei district of Changzhou, which is a newly developed urban area. The form of the tower is inspired by the 1,300 year-old Tianning Temple, which is the most famous historic building in Changzhou. The project intends to represent a pagoda, which in traditional Chinese culture is thought to bring good fortune. The lower part of the main tower is office space, and the higher section is a Marriott hotel. The top of the tower consists of luxury apartments and an observatory.

The complex also contains other functions, including radio and TV broadcasting, retail, and a theater. To avoid the negative impact of a superblock on the surrounding area, the tower's podium is divided into smaller blocks, forming internal streets that can be navigated at pedestrian scale. The placement of the tower on the site is intended to maximize its contribution to the urban landscape, and to reduce the blockage of light to the residential buildings behind.

Completion Date: August 2014
Height: 256 m (840 ft)
Stories: 58
Area: 89,767 sq m (966,244 sq ft)
Use: Residential/Hotel/Office
Owner: Changzhou Broadcasting Station
Developer: Changzhou Radio and TV Realty Company, Ltd.
Architect: Shanghai Institute of Architectural Design & Research, Co., Ltd.
Structural Engineer: Shanghai Institute of Architectural Design & Research, Co., Ltd.
MEP Engineer: Shanghai Institute of Architectural Design & Research, Co., Ltd.
Main Contractor: China Construction Third Engineering Bureau Co., Ltd.

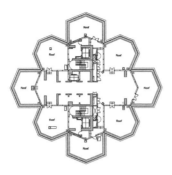

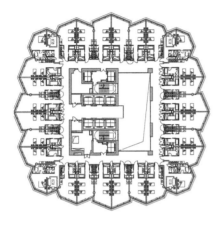

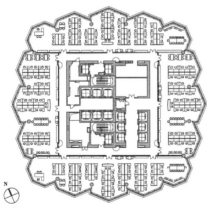

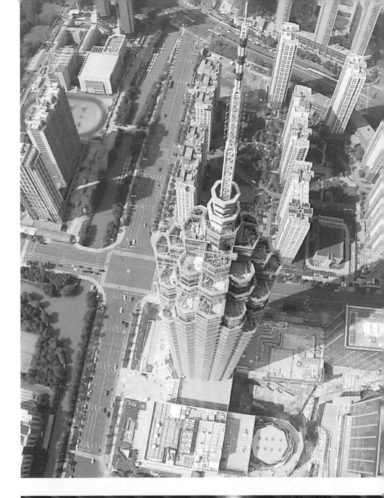

Opposite: Overall view of tower from Longjin Road

Above: Typical floor plans – observatory level (top), typical hotel level (middle) and typical office level (bottom)

Top Right: Aerial view of tower

Bottom Right: Main broadcasting and office room

Nominee
Best Tall Building Asia & Australasia

China Merchants Tower
Shenzhen, China

China Merchants Tower anchors the Woods Park master plan, located in the Nanshan District of Shenzhen, providing new office and select retail space. The tower bows out from its base and tapers and elongates up its height, creating a striking form. The chamfer and reveal in its massing articulate its façade, while incorporated facets emphasize the tower's verticality and slender proportion. The tower's shape also allows for the distribution of exterior notches, supporting balconies.

The tower features a low-E, unitized glass curtain wall clad in a system of horizontal glass fins and vertical aluminum struts. To control the impact of Shenzhen's hot and sunny climate, the façade's fins have been closely spaced together in order to reduce solar gain. The fins give the façade a fine-grain texture and balance the slenderness of the tower's verticality. They also refract light at night, illuminating the form.

Completion Date: December 2013
Height: 211 m (ft)
Stories: 38
Area: 107,000 sq m (sq ft)
Use: Office/Retail
Owner/Developer: China Merchants Real Estate Shenzhen Co., Ltd.
Architect: Skidmore, Owings & Merrill (design); The Architectural Design & Research Institute of Guangdong Province (architect of record)
Structural Engineer: Skidmore, Owings & Merrill; The Architectural Design & Research Institute of Guangdong Province
MEP Engineer: Skidmore, Owings & Merrill; The Architectural Design & Research Institute of Guangdong Province
Main Contractor: The Second Construction Engineering Co., Ltd. (CSEC)
Other Consultants: Arup (façade); BLVD (interiors); Brandston Partnership, Inc. (lighting); EDAW (landscape); Fortune Consultants, Ltd. (vertical transportation); Guangdong Provincial Academy of Building Research (wind); Hua Nan Technology University (energy concept); The Architectural Design & Research Institute of Guangdong Province (fire); Rider Levett Bucknall (cost)

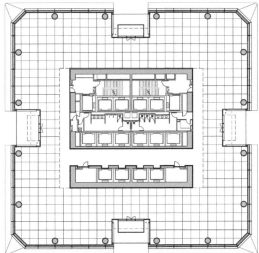

Opposite: Overall view of tower from south

Top Left: Typical floor plan

Bottom Left: Tower lobby from approach

Top Right: Texture of façade from tower base at night

Bottom Right: Façade detail

Nominee
Best Tall Building Asia & Australasia

Fake Hills
Beihai, China

Despite its apparently tongue-in-cheek name, Fake Hills presents a genuine hybrid experience between high-rise life and a seaside park. Combining the best of both typologies, providing for oceanfront views, while giving access to green, open-air terraces along the entire roofline, its profile is a carefully constructed datum line created to reflect China's flowing mountains as they rise above the dense fog below. An arched entryway and cut-out hole provide visual and spatial continuity between the oceanfront and the valley behind the building.

This development is located in the coastal city of Beihai, China, on a 800-meter-long, narrow oceanfront site. The fundamental geometry of the scheme combines two common yet opposite architecture typologies; the high-rise and the "groundscraper," producing an undulating building typology, resulting in the form of a hill. The geometry of the architecture maximizes potential views for the residents; the continuous platform along the roof becomes the public spaces, with gardens, tennis courts, and swimming pools on top of the man-made hills.

Completion Date: 2014
Height: 106 m (348 ft)
Stories: 33
Area: 492,369 sq m (5,299,816 sq ft)
Use: Residential
Developer: Beihai Xinpinguangyang Real Estate Development Co., Ltd.
Architect: MAD Architects
Structural Engineer: Jiang Architects and Engineers
MEP Engineer: Jiang Architects and Engineers

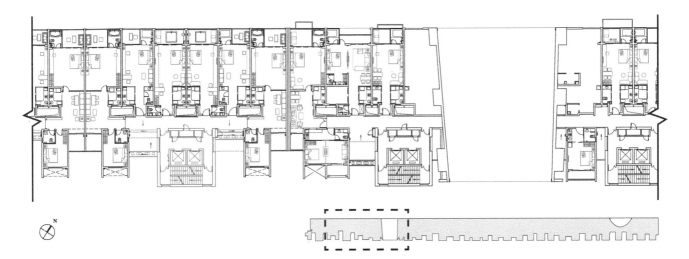

Opposite Top: Overall view from southwest

Opposite Bottom: View of public space at tower base

Top: Partial typical floor plan with key

Above: View of public space from balcony

Right: Building entrance, cut-out and balcony detail

Nominee
Best Tall Building Asia & Australasia

Guangzhou Circle
Guangzhou, China

The Guangzhou Circle acts as an "urban logo" that serves as a landmark, in the same way that ideograms are used in Chinese writing, instead of an alphabet. As such, it may seem indecipherable to those conditioned by Western skyscraper typologies. The "Bi" disc is one of China's most enduring symbols, with a history going back 5,000 years. Here, set alongside the Zhujiang River, the reflection of the disc-shaped structure forms a figure "8" in the water, also an enduring symbol of good fortune.

The Circle plays a strong role in the landscape. It is the south gate of Guangzhou, and by extension, for all of China, as the city is a terminus for ferry boats and high-speed rail. While much of industrial China is characterized by anonymity and repetition, here a kind of "cultural sustainability" is at work. People can recognize in this arresting form the essence of their territory and traditions.

Completion Date: December 2013
Height: 138 m (453 ft)
Stories: 33
Area: 83,000 sq m (893,405 sq ft)
Use: Office
Owner: Guang Dong Hong Da Xing Ye Group
Developer: Guang Dong Hong He Construction, Ltd.
Architect: AMproject (design); South China University of Technology (architect of record)
Structural Engineer: SIGGMA (design); South China University of Technology (engineer of record)
MEP Engineer: South China University of Technology
Main Contractor: Guang Dong Hong He Construction, Ltd.
Other Consultants: Shen Zhen Jian Hong Da Construction, Ltd. (interior); Shen Zhen Rui Hua Construction, Ltd. (façade)

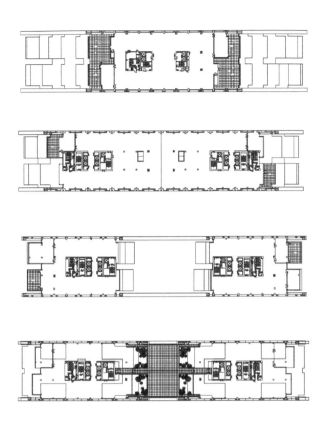

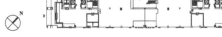

Opposite: View of the tower from street level

Above: Floor Plans (top to bottom) – level 33, level 27, level 18, level 10, level 7, level 1

Top Right: Overall view of river reflection

Bottom Right: Detail of façade and structure

Habitat
Melbourne, Australia

Habitat is a project fundamentally informed by its context. The tower realizes the latent potential of a small site adjacent to an elevated section of the West Gate freeway and gives physical expression to the site's challenging constraints. Small windows punched into precast façades face the freeway. When viewed from a distance the windows, which are subtlety different in size, blend into a singular, city-scale gesture, lending the building a strong identity amidst the anonymity of Southbank's high-rise apartment buildings, referencing the sound waves that the building has been designed to negate.

In recognition of the site constraints, the building adopts an innovative approach to the provision of resident amenity. The 147 one- and two-bedroom apartments are clustered around large three-story sky gardens. Opening to the east, the communal gardens afford expansive views over central Melbourne and eastern suburbs while taking advantage of the most climatically benign of the building's aspects.

Completion Date: June 2013
Height: 109 m (358 ft)
Stories: 35
Area: 12,574 sq m (135,345 sq ft)
Use: Residential
Owner/Developer: Vicland Corporation Pty, Ltd.
Architect: ROTHELOWMAN
Structural Engineer: Bonacci Group
MEP Engineer: Jeff Bryar & Associates
Main Contractor: Maxcon Pty, Ltd.
Other Consultants: Cardno (traffic); Contour Consultants Australia (urban planner); McKenzie Group (quantity surveyor); MEL Consultants Pty., Ltd. (wind); Renzo Tonin & Associates (acoustics)

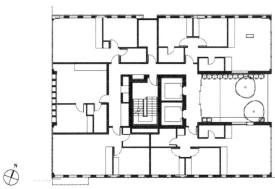

Opposite: Overall view of tower from northeast

Top Left: Detail view of glass façade and green slab balconies

Bottom Left: Typical floor plan

Top Right: View of building from south

Bottom Right: Interior view of bedroom with punched openings

Nominee
Best Tall Building Asia & Australasia

Jinao Tower
Nanjing, China

The Hexi District of Nanjing is in the midst of transforming from a rural area to a vibrant business hub and civic center. Jinao Tower is an anchor and gateway to the new district, serving as an icon of the area's new urbanity. This next-generation commercial and hotel tower maximizes performance, efficiency, and occupant experience. Its distinctive external bracing system creates more efficient lateral support – requiring less steel and an overall 20 percent reduction in building material.

The design of the tower is rooted in the notion of developing this parcel of land and the neighboring parcel to serve both as a gateway to the new district and as a symbol of the vitality of the district. Its faceted form is derived from the juxtaposition of the innovative double-skin façade that creates solar shading and an insulating chamber between the building envelope and the occupied space. Vented openings in the outermost curtain wall allow wind pressure to draw built-up heat out of the cavity, lowering temperatures along the inner exterior wall.

Completion Date: March 2014
Height: 232 m (761 ft)
Stories: 56
Area: 168,000 sq m (1,808,337 sq ft)
Use: Hotel/Office
Owner/Developer: Jiangsu Goldenland Real Estate Development (Group) Co., Ltd.
Architect: Skidmore, Owings & Merrill
Structural Engineer: Skidmore, Owings & Merrill
MEP Engineer: WSP Group
Main Contractor: Wuhan Construction Engineering Group Co., Ltd.
Other Consultants: Brandston Partnership, Inc. (lighting); CS Caulking Co., Inc. (façade maintenance); Edgett Willams Consulting Group, Inc. (vertical transportation); SWA Group (landscape); WSP Group (fire)

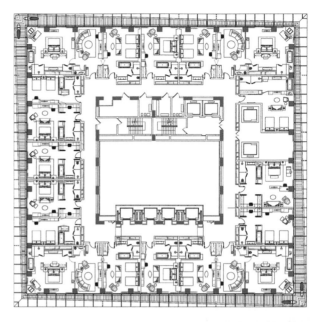

Opposite: Overall view of tower

Above: Section showing double-skin façade and structure

Top Right: Typical hotel level floor plan

Bottom Right: Hotel atrium in upper half of tower

Kent Vale

Singapore

An addition to the existing faculty apartments on the campus of the National University of Singapore, Kent Vale was designed to be an iconic gateway building, where the realms of domesticity and civic monumentality would meet. The challenge of the brief lay in the melding of two distinct uses: first, as a new administrative center, where architecture should present itself as a statement of the client's vision for a new campus ideal; second, as a quiet oasis for residents to interact and enjoy.

This tropical high-rise is notable for its adventurous use of high-rise greenery. The façade, though intricate, is not merely decorative. Solar gain through the façade is minimized via optimum building orientation, strategic sun-shading by modular green walls, balconies, and screens on the east- and west-facing façades, and horizontal sun-shading on all façades. The design of Kent Vale presents the culmination of sustained analytical design, melding function, sustainability, and constructability in a highly resolved architectural form.

Completion Date: September 2012
Height: Tower 1: 91 m (299 ft); Tower 2: 87 m (285 ft); Tower 3: 87 m (285 ft)
Stories: Tower 1: 25; Tower 2: 24; Tower 3: 24
Area: 49,910 sq m (537,227 sq ft)
Use: Residential
Owner/Developer: National University of Singapore
Architect: MKPL Architects Pte., Ltd.
Structural Engineer: KTP Consultants Pte., Ltd.
MEP Engineer: J. Roger Preston Group
Main Contractor: Tiong Seng Constractors Pte., Ltd.
Other Consultants: AECOM / Davis Langdon (quantity surveyor); Lighting Planners Associates (S) Pte Ltd (lighting); Sitetectonix Pte., Ltd. (landscape)

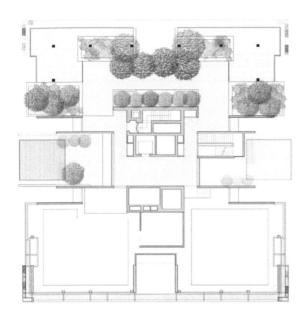

Opposite Top: Overall view of towers from east

Opposite Bottom: Interior façade shading detail

Above: Floor plans – roof terrace (top) and typical floor (bottom)

Top Right: View looking down at landscaping from balcony

Bottom Right: Green wall façade detail

Nominee
Best Tall Building Asia & Australasia

L'Avenue
Shanghai, China

Located in Shanghai's Chang Ning District, L'Avenue is a commercial tower development that includes an office tower, a four-story retail podium, and a four-level basement. Built as a flagship project for a prestigious retail brand, the building's unusual shape renders it a landmark in the district. The unique architectural form conveys the subtle fluidity of the overall design from the tower to the transitional, 60-meter-span skylight, finishing at the curving podium structure. This shell structure of 175-millimeter reinforced concrete with internal stiffeners, finished with glazed ceramic tiles, captures the subtle sparkles emanating from this dynamic, freeform structure.

The freeform approach used in the design of the building's curvaceous and twisting shape immediately conjures images of an elegant evening gown, while the integration of architectural fins and lighting within the building envelope introduces the concept of a "waterfall" into the lighting design. The design also has environmental credentials – the project received LEED Gold certification in February 2013.

Completion Date: July 2012
Height: 134 m (440 ft)
Stories: 28
Area: 64,457 sq m (693,809 sq ft)
Use: Office/Retail
Owner/Developer: Shanghai Luxchina Property Development Co., Ltd.
Architect: Leigh and Orange Ltd. (lead consultant/architect); Jun Akoi & Associates (concept architect)
Structural Engineer: WSP Group
MEP Engineer: TAP Consulting Engineers, Ltd.
Project Manager: STDM
Main Contractor: China Construction Eighth Engineering Division
Other Consultants: HS&A (façade); MVA Transportation (traffic); WSP Group (sustainability)

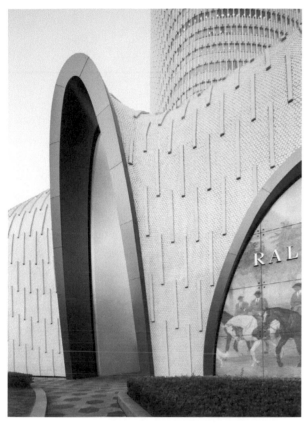

Opposite Top: Overall view from northwest

Opposite Bottom: Aerial view of atrium skylight

Top Left: Entrance to shopping mall

Bottom Left: Central atrium

Top Right: Façade lighting detail at night

Bottom Right: Typical section

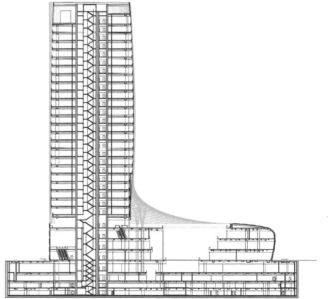

Nominee
Best Tall Building Asia & Australasia

OLIV
Hong Kong, China

OLIV is a retail building within the busy Causeway Bay shopping district in Hong Kong. Sitting on a small site, the narrow building is nonetheless able to offer a floor-to-floor height of 5 meters. All lifts open directly into tenant spaces, without passing through lobbies or corridors. The design is inspired by the olive tree, after which the building was named. The floor plate changes slightly at each level, as the building moves upward, giving it a twisting and turning profile.

The exterior is wrapped with a "triple skin," with an interior orthogonal layer of sub-frame and mullions, a dark grey glass curtain wall layer, and an "organic" white aluminum cladding. This layer contains the "olive tree knots" which turn into radiating "stars" – the signature and logo of OLIV. Every piece of the aluminum cladding is different, due to the irregular form of the building. Each was prefabricated off-site, calibrated to fit perfectly onto the concrete structure.

Completion Date: January 2014
Height: 136 m (445 ft)
Stories: 25
Area: 4,305 sq m (46,339 sq ft)
Use: Retail
Owner/Developer: Benway Limited
Architect: THEO TEXTURE (design); Studio Raymond Chau Architecture Limited (architect of record)
Structural Engineer: Wong & Cheng Consultants Engineers Limited
MEP Engineer: Trustful Engineering & Construction Co., Ltd.
Main Contractor: Trustful Engineering & Construction Co., Ltd.
Other Consultants: Harbour Century Limited (quantity surveyor)

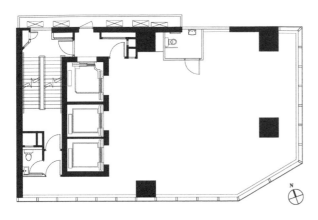

Opposite: Overall view from street

Top Left: 11th floor sky garden floor plan

Bottom Left: Detailed view of the façade

Top Right: Façade illumination at night

Bottom Right: Public space on the 11th floor sky garden

Nominee
Best Tall Building Asia & Australasia

Shanghai Arch
Shanghai, China

The name for the Shanghai Arch mixed-use complex was inspired by the "portal" design of its signature office building. As the first of three components that make up the complex, the office building acts as a gateway into the other project components, as well as to the overall Hongqiao Expansion Business Zone.

Rising up as two separate towers, the building is joined at the 23rd floor by a seven-story skybridge that creates a dynamic "gate" into the pedestrian-friendly retail promenade and offers larger floor plates on the building's upper levels. The east wall is subtly curved with overlapping glass panels, and the east and west elevations feature a double-skin façade for enhanced thermal performance. The glazing at street level blurs the distinction between inside and out, complemented by the flow of the plaza landscaping and water features that "pass through" the office building and connect it to the surrounding context.

Completion Date: September 2014
Height: 144 m (472 ft)
Stories: 29
Area: 262,476 sq m (2,825,268 sq ft)
Use: Office
Owner/Developer: Shanghai Jin Hong Qiao International Property Co., Ltd.
Architect: John Portman & Associates (design); East China Architectural Design & Research Institute (architect of record)
Structural Engineer: John Portman & Associates; East China Architectural Design & Research Institute
MEP Engineer: Newcomb & Boyd; East China Architectural Design & Research Institute
Main Contractor: Shanghai No. 7 Construction Co., Ltd
Other Consultants: ALT Cladding (façade); Arnold Associates (landscape); Fortune Consultants, Ltd. (vertical transportation)

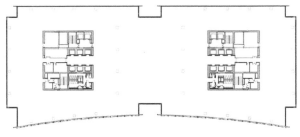

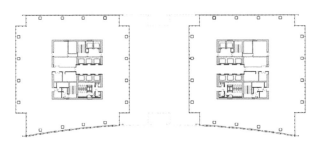

Opposite Top: Overall view of tower from east

Opposite Bottom: View of lobby

Top: Looking up at the tower bridge

Above: Façade detail

Right: Floor plans – bridge plan (top) and mid-level plan (bottom)

Nominee
Best Tall Building Asia & Australasia

Xiamen Financial Centre
Xiamen, China

The Xiamen Financial Centre is the most dominant tower in the new central business district of Xiamen. The iconic sculptural form of the tower creates a striking image, while its folded façades also help to break down its mass in a tight urban site. The system of folding surfaces and joint lines was carried through into different building elements and functional spaces, such as the entrance canopy, the elevated lobby, and the car park ramp shelter.

The glass "iceberg" tower has an elevated lobby, taking the occupant to the escalator alongside the civic square to the south, culminating in a direct eastern view to the sea. By elevating the lobby, a significant amount of ground floor area is dedicated to public space at the city scale. With a sky clubhouse on the 15th floor and an observation deck on the top floor, occupants are able to enjoy the panoramic sea view from different levels.

Completion Date: July 2013
Height: 213 m (699 ft)
Stories: 49
Area: 139,502 sq m (1,501,587 sq ft)
Use: Office
Owner/Developer: Xiamen Land Development Company
Architect: Gravity Partnership (design); Xiamen BIAD Architectural Design Co., Ltd. (architect of record)
Structural Engineer: Beijing Institute of Architectural Design; Xiamen BIAD Architectural Design Co., Ltd.
MEP Engineer: Xiamen BIAD Architectural Design Co., Ltd.
Main Contractor: China State Construction Engineering Corporation
Other Consultants: Arup (façade); BIAD Lighting Design Studio (lighting); Gravity Green Ltd. (landscape)

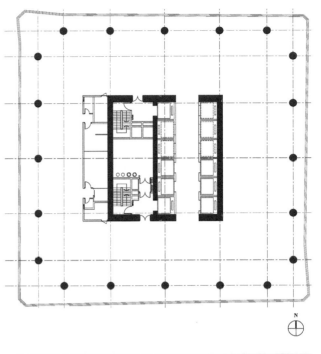

Opposite: Overall view of tower from southwest

Top Left: View of tower looking up from entrance

Bottom Left: Entrance from street

Top Right: Mid-zone floor plan

Bottom Right: Elevated lobby with structural detail

ASE Centre Chongqing R2
Chongqing, China

Completion Date: 2013
Height: 204 m (669 ft)
Stories: 58
Area: 51,568 sq m (555,073 sq ft)
Use: Residential
Owner/Developer: ASE Group
Architect: Dennis Lau & Ng Chun Man Architects & Engineers
Structural Engineer: CISDI
MEP Engineer: CISDI
Main Contractor: ASE Group
Other Consultants: Earth Asia (landscape)

Asia Square
Singapore

Completion Date: November 2013
Height: Tower 1: 229 m (751 ft); Tower 2: 221 m (727 ft)
Stories: Tower 1: 43; Tower 2: 46
Total Area: 246,476 sq m (2,653,046 sq ft)
Use: Hotel/Office
Owner: Asia Square Tower Pte., Ltd.
Developer: BlackRock Property Singapore Ptd., Ltd.
Architect: Denton Corker Marshall (design); Architects 61 (architect of record)
Structural Engineer: AECOM
MEP Engineer: AECOM
Main Contractor: Hyundai Engineering & Construction
Other Consultants: Acviron Acoustics Consultants Pte Ltd (acoustics); ALT Cladding (façade); Brandston Partnership, Inc. (lighting); Building System and Diagnostics (environmental); Earthscape Concepts (landscape); FBEYE International (interiors); Integrated Building Consultants (fire); Northcroft Lim Consultants (quantity surveyor)

China Resources Building
Hong Kong, China

Completion Date: Original: 1983; Renovation: January 2013
Height: 178 m (584 ft)
Stories: 50
Area: 99,000 sq m (1,065,627 sq ft)
Use: Office
Owner/Developer: China Resources Property, Ltd.
Architect: Ronald Lu & Partners
Structural Engineer: Siu Yin Wai & Associates, Ltd.
MEP Engineer: Talent Mechanical and Electrical Engineers, Ltd.
Main Contractor: CR Construction Co., Limited
Other Consultants: Alpha Consulting, Ltd. (façade); CL3 Architects, Ltd. (interiors); Langdon & Seah (quantity surveyor); Arup (sustainability); Spectrum Design and Associates (lighting); Urbis (landscape)

DBS Bank Tower
Jakarta, Indonesia

Completion Date: March 2013
Height: 194 m (636 ft)
Stories: 40
Area: 73,000 sq m (785,765 sq ft)
Use: Office
Owner: Ciputra Group
Developer: PT Ciputra Property, Tbk.
Architect: RTKL (design); PT Perentjana Djaja (architect of record)
Structural Engineer: BECA Group; Wiratman & Associates
MEP Engineer: BECA Group, Ltd.; Arkonin
Main Contractor: Jaya Konstruksi; Tata; Nusa Raya Cipta
Other Consultants: CCW Associates Pte., Ltd. (acoustics); Belt Collins & Associates (landscape); Kaplan Gehring McCarrol Architectural Lighting, Inc. (lighting); Hadi Komara (lighting); Meinhardt (façade); Pamintori Cipta (traffic); Reynolds & Partners (quantity surveyor); Windtech Consultants (wind)

Fortune Plaza Phase III
Beijing, China

Completion Date: 2014
Height: 267 m (876 ft)
Stories: 61
Area: 151,585 sq m (1,631,647 sq ft)
Use: Office
Owner/Developer: Xiang Jiang Xing Li Estates Development, Ltd.
Architect: P & T Group (design); CERI, Ltd. (architect of record)
Structural Engineer: Arup; CERI, Ltd.
MEP Engineer: Parsons Brinckerhoff; CERI, Ltd.
Main Contractor: HKI China Land, Ltd.
Other Consultants: ACLA (landscape); EMSI (LEED); Schmidlin (façade); WT Partnership (cost)

Infinity
Brisbane, Australia

Completion Date: July 2013
Height: 249 m (817 ft)
Stories: 81
Area: 59,136 sq m (636,535 sq ft)
Use: Residential
Owner/Developer: Meriton Group
Architect: DBI Design Pty., Ltd.
Structural Engineer: Bonacci Group
MEP Engineer: ADG; Superior Air; Suncorp Electrical
Main Contractor: Pryme
Other Consultants: Acoustic Logic (acoustic); Auscoast Fire Services (fire); Certis (certifier); Cityplan Services (planning); Golder & Associates (geotechnical); Halcrow MWT (traffic); HCE Engineers (civil); MEL Consultants Pty., Ltd. (wind)

Jinling Hotel Asia Pacific Tower

Nanjing, China

Completion Date: May 2014
Height: 242 m (794 ft)
Stories: 57
Area: 123,361 sq m (1,327,847 sq ft)
Use: Hotel/Office
Owner: Jinling Hotel Corporation, Ltd.
Developer: New Jinling Hotel Limited Company
Architect: P & T Group (design); Jiangsu Provincial Architectural D&R Institute, Ltd. (architect of record)
Structural Engineer: P & T Group, Ltd.
MEP Engineer: Jiangsu Provincial Architectural D&R Institute, Ltd.
Main Contractor: China State Construction Engineering Corporation
Other Consultants: Campbell Shillinglaw Lau, Ltd. (acoustics); Chhada Siembieda Leung, Ltd. (interior); Meinhardt (façade); Nanjing Institute of Landscape Architecture Design & Planning, Ltd. (landscape); Shanghai Citelum Lighting Design Co., Ltd. (lighting); Watermark Associates (way finding)

One AIA Financial Center

Foshan, China

Completion Date: January 2013
Height: 141 m (463 ft)
Stories: 28
Area: 52,002 sq m (559,745 sq ft)
Use: Office
Owner: AIA Group Limited
Developer: Foshan Main Forum Real Estate Development Company Limited
Architect: Aedas
Structural Engineer: Meinhardt
MEP Engineer: J. Roger Preston Group
Main Contractor: China Construction Third Engineering Bureau Co., Ltd.
Other Consultants: Aurecon (façade); Campbell Shillinglaw Lau, Ltd. (acoustics); Hyder Consulting (LEED); Urbis (landscape); WT Partnership (quantity surveyor)

RMIT Swanston Academic Building

Melbourne, Australia

Completion Date: June 2013
Height: 52 m (171 ft)
Stories: 12
Area: 34,000 sq m (365,793 sq ft)
Use: Education
Owner/Developer: RMIT University
Architect: Lyons
Structural Engineer: Bonacci Group
MEP Engineer: AECOM
Project Manager: DCWC
Main Contractor: Brookfield Multiplex
Other Consultants: AECOM (vertical transportation); Architecture & Access (access); Bonacci Group (civil); Meinhardt (façade); MEL Consultants Pty., Ltd. (wind); PLP (certifier); Rush Wright Associates (landscape); Wilde & Woollard (quantity surveyor)

The Capital

Mumbai, India

Completion Date: January 2013
Height: 78 m (256 ft)
Stories: 19
Area: 62,563 sq m (673,423 sq ft)
Use: Office
Owner/Developer: The Wadhwa Group
Architect: James Law Cybertecture International (design); The Wadhwa Group (architect of record)
Structural Engineer: Mahimtura Consultants, Pvt., Ltd.
MEP Engineer: MEP Consulting Pty., Inc.
Main Contractor: The Wadhwa Group
Other Consultants: Bo Steiber Lighting Design (lighting); HBO+EMTB Group (landscape); Meinhardt (façade); Wohr Parking Systems Pvt., Ltd. (parking)

The Gloucester
Hong Kong, China

Completion Date: January 2013
Height: 164 m (539 ft)
Stories: 41
Area: 10,739 sq m (115,594 sq ft)
Use: Residential
Owner/Developer: Henderson Land Development Company Limited
Architect: Dennis Lau & Ng Chun Man Architects & Engineers
Structural Engineer: Stephen Cheng Consulting Engineers Limited
MEP Engineer: J. Roger Preston Group
Main Contractor: Heng Lai Construction Company Limited
Other Consultants: AB Concept Limited (façade); Adrian Norman Limited (landscape); AEC Limited (environmental); J. Roger Preston Group (fire); MVA Transportation (traffic); Planning & Management Consultants (traffic)

The Pakubuwono Signature
Jakarta, Indonesia

Completion Date: July 2014
Height: 252 m (827 ft)
Stories: 50
Area: 75,714 sq m (814,979 sq ft)
Use: Residential
Owner: The Pakubuwono Development
Developer: The Pakubuwono Development; Mandiri Eka Abadi
Architect: Airmas Asri
Structural Engineer: Davy Sukamta & Partners
MEP Engineer: Hantaran Prima Mandiri
Main Contractor: ACSET
Project Manager: Jaya CM
Other Consultants: Langdon & Seah (quantity surveyor); Insada Integrated Design Team (interiors); Lumina (lighting)

Europe

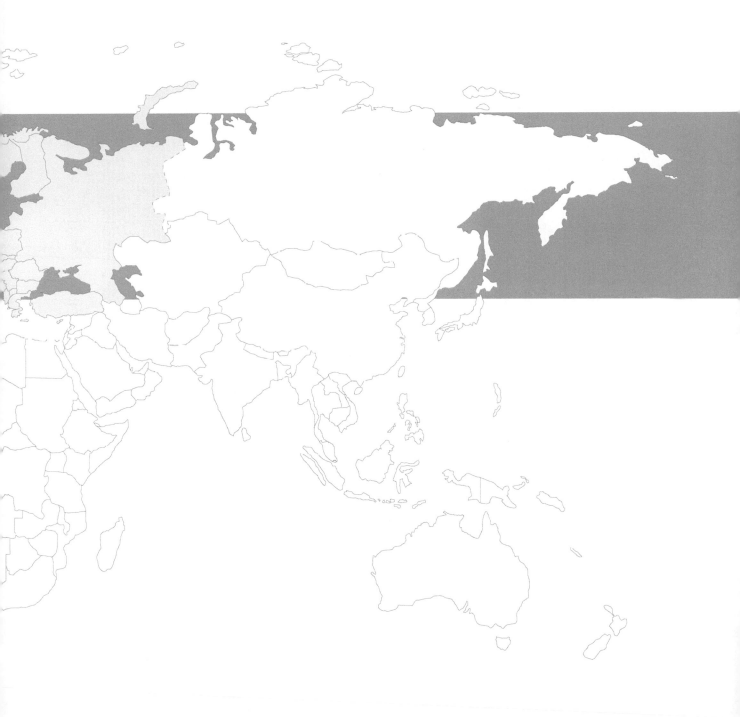

De Rotterdam
Rotterdam, Netherlands

"This scheme demonstrates big memorable architecture that intelligently addresses formal and organizational relationships with the city."

Sir Terry Farrell, Juror, Farrells

Completion Date: November 2013
Height: 151 m (495 ft)
Stories: 45
Area: 162,000 sq m (1,743,753 sq ft)
Use: Residential; Office; Office/Hotel
Developer: MAB; OVG Projectontwikkeling
Architect: Office for Metropolitan Architecture
Structural Engineer: Corsmit Raadgevende Ingenieurs
MEP Engineer: Techniplan Adviseurs; Valstar Simonis
Project Manager: De Rotterdam CV; DVP
Main Contractor: Zublin
Other Consultants: ABT Delft (code); Arup (structural advisor at schematic stage); DGMR Raadgevende Ingenieurs (acoustics, fire, wind); Permasteelisa Group (façade); TGM (façade)

De Rotterdam is conceived as a vertical city: three interconnected mixed-use towers accommodate offices, apartments, a hotel, conference facilities, shops, restaurants, and cafés. Unlike a typical tall building, where the programs are stacked one on top of another, in De Rotterdam, the functions come in side by side (see floor plans).

The towers are part of the ongoing redevelopment of the old harbor district of Wilhelminapier, next to the Erasmus Bridge, and aim to reinstate the vibrant urban activity – trade, transport, leisure – once familiar to the neighborhood. De Rotterdam is named after one of the ships on the Holland America Line, which departed from the Wilhelminapier in decades past, carrying thousands of Europeans emigrating to the US.

The project consists of three connected towers that appear as stacked and shifted volumes upon a base plinth. The façade design has been kept neutral and transparent; the dynamic appearance of the building is determined by the varied day cycles of the different programs. The deep mullions allow the glazed façades to appear more open or closed, depending on the perspective.

De Rotterdam is an exercise in formal interpretation that is at once reminiscent of an imported mid-20th-century American skyscraper, but epitomizes the off-center experimentalism of modern Dutch art of

the foregoing century. The nighttime twinkling of the lights indicating different programs lends dynamism and contributes to the humanization of the monoliths. It is as if the moai of Easter Island were constantly craning their necks and raising their eyebrows at the change all around.

At ground floor level, a ceiling height of 8.5 meters ensures a smooth transition between exterior and interior, while loading areas are kept, as much as

possible, to other levels. The street-side entrance zone is separated from the waterfront restaurant zone by only a relatively small core, and a large central lobby ensures that a visual connection between street and waterfront remains. The public program on the upper floors of the plinth meets the ground floor in a large atrium; this void extends externally between the low-rise tower volumes to 85 meters' height. Once within the towers, views between the hotel, office and residential volumes continue the theme of transparency.

The three towers reach 150 meters high, with a gross floor area of approximately 162,000 square meters, making De Rotterdam the largest building in the Netherlands. The architectural concept produces more than sheer size: urban density and diversity – both in the program and the form – are the guiding principles of the project. De Rotterdam's stacked towers are arranged in a subtly irregular cluster that refuses to resolve into a singular form, and produces intriguing new views from different perspectives.

Jury Statement

De Rotterdam subverts accepted notions of how a skyscraper is supposed to behave. While the collective massing suggests a refined and simple monolith, the slightest change of perspective reveals secondary and tertiary complexities. Sunsets cascade through the small gaps between the offset upper volumes, as if the building is some kind of ancient timekeeping device. In certain lights, the towers are shimmering and immutable; at dusk, they are translucent, revealing the K-braces and other structural interventions required to achieve the impression of shifted solidity. The translucence also lends levity to the otherwise massive building, re-confirming its daytime appearance as the sail of a large ship while re-inventing it at night as a box lantern at the harbor's end.

The towers take a straightforward approach to separating distinct programs into distinct blocks, yet attain an interesting diversity through their slightly irregular forms and interconnectedness. De Rotterdam provokes thoughts about separateness, context, structure, and façade, all at once. It will not be surprising if future generations make explicit pilgrimages to view it from different angles in different seasons, and from land as well as sea.

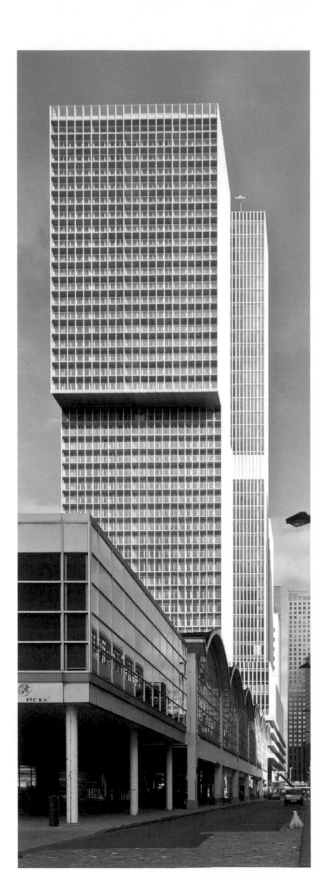

The various programs of this urban complex are organized into distinct blocks, providing both clarity and synergy: residents and office workers alike can use the fitness facilities, restaurants, and conference rooms of the hotel. These private users of the building have contact with the general public on the ground floor, with its waterfront cafés. The lobbies for the offices, hotel, and apartments are located in the plinth – a long elevated hall that serves as a general traffic hub for De Rotterdam's wide variety of users.

Comprehensive building management, including an energy monitoring system, has been employed to ensure maximum efficiency throughout the project. For the purpose of energy supply, a collective generation system was developed, which feeds all the functions in the building. Power is generated via district heating and co-generation with biofuel, while water from the adjacent Maas river is used for cooling. The temperature system additionally uses low-temperature heating and high-temperature cooling, heat exchangers for heat recovery ventilation, and fan speed to control air handing.

Maximum use of daylight is supplemented by efficient artificial lighting, using high-efficiency reflectors. Appropriate lighting methods have been selected for the various functions, with automatic daylight and motion control in the office areas, and LEDs in the public zones. Sustainability is further improved with water-saving taps and reservoirs, and efficient elevators using energy recovery

Housing

Offices

Offices

Offices

Offices

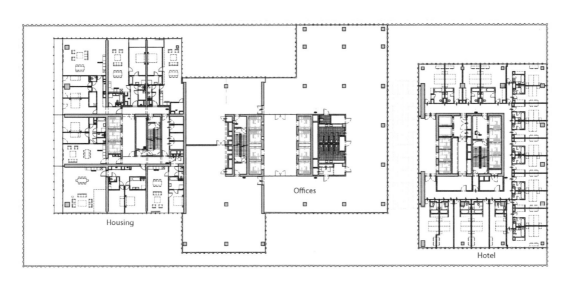

Housing

Offices

Hotel

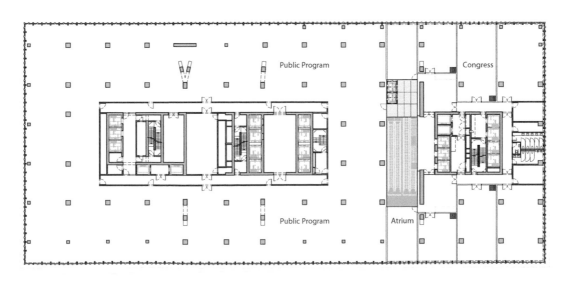

Public Program

Congress

Public Program

Atrium

DC Tower

Vienna, Austria

Completion Date: February 2014
Height: 220 m (722 ft)
Stories: 60
Area: 93,500 sq m (1,006,426 sq ft)
Use: Residential/Office/Hotel
Owner/Developer: Wiener Entwicklungsgesellschaft für den Donauraum AG
Architect: Dominique Perrault Architecture (design); Hoffmann-Janz Architekten (architect of record)
Structural Engineer: Bollinger + Grohmann; Gmeiner Haferl Zivilingenieure ZT GmbH
MEP Engineer: AXIS Ingenieursleistungen ZT GmbH; Eipeldauer & Partner GmbH; ZFG-Projekt
Main Contractor: Max Bögl Bauunternehmung GmbH & Co; STRABAG AG
Other Consultants: Dr. Pfeiler GmbH; ELIN GmbH & co KG; OK Osadnik & Kamienski GmbH; Prüfstelle für Brandschutztechnik (fire); Stahlform Baustahlbearbeitungs GmbH; Vermessung Angst ZT GmbH; Wacker Ingenieure (wind); Werner Sobek Group (façade); YIT Austria GmbH

> *"The ribbons of the main façade of DC Tower give the building a kinetic energy and dynamism that hints at an ability of the building to actually move."*
>
> Antony Wood, Juror, CTBUH

Austria's tallest building, DC Tower, has become an invaluable landmark of the Donau-city in Vienna. The 220-meter building can be compared to an entirely new urban district with a diverse range of functions: offices, a four star hotel, apartments, a sky bar, a public open space, restaurants, and a fitness center. A subtle game of flat and folded façades affords the glass and steel tower a sensual identity.

The façade folds give the tower a liquid, immaterial character, a malleability constantly adapting the light, a reflection or an event. The folds contrast with the no-nonsense rigor of the other three façades, creating a tension that electrifies the public space at the tower base. Dancing on its platform, the tower is slightly oriented toward the river to open a dialogue with the rest of the city, turning its back on no one, neither the historic nor the new Vienna.

The aim of the designers was to get the basic horizontality of the city and the public space to

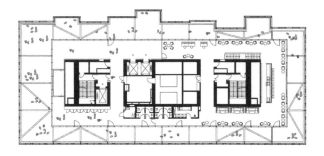

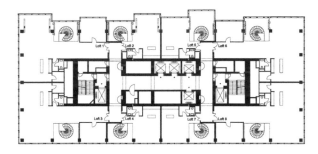

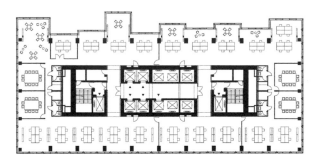

coincide with vertical trajectories. On the back façade, the public space rises from the level of the esplanade in a series of staggered steps to reach the ground reference plane. This structuring of topography launches the tower and creates a spatial interface accessible to all, making the occurrence of such a physical object both possible and acceptable. On the other three façades, 54 metallic umbrellas gradually rise from the ground on the approach, softening the hard edges of the project and blending city and movement into the tower's future. Important work on neighborhood fringes remains to be done to reveal the geographic features of this urban landscape and take better advantage of the river bank.

With this first substantial tower, the city of Vienna has demonstrated that the punctual and controlled emergence of high-rises can participate in creating the city and produce contemporary, economical, high-energy performance, mixed-use buildings, adapted to metropolitan business requirements and lifestyles.

In each office, openings provide natural fresh air. The curtain walls are composed of three different glass layers, so as to provide solar protection.

Active floors, used to a large extent in the tower, minimize energy consumption by reducing the volume of air flow to exactly match the thermodynamic heat load. Water recycled from the adjacent Danube River is used for the cooling process.

"Placing a pronounced and jagged exclamation mark on the skyline of a city as restrained and refined as Vienna is an act of bravado, and may come to signify a moment of change in the city's self-conception."

Saskia Sassen, Juror, Columbia University

Previous Spread

Left: Overall view of tower

Right: Façade detail from tower base

Current Spread

Left: Floor plans – Level 59 sky garden (top), level 53 lofts (middle) and level 25 offices (bottom)

Opposite Left: Tower base with plaza umbrella structures

Opposite Top Right: Rooftop terrace

Opposite Bottom Right: View of lobby

With its confident presence, the DC Tower affirms the culmination of a relationship between Vienna and the Danube that has historically been tentative. The flood-prone Donau City area has gradually been developed since the 1970s, but lacked presence somewhat. DC Tower's onyx form and tessellating façade are like a lightning bolt striking a stake in the ground – Donau City is Vienna's sister, but has its own identity. Now the metropolis has a solid, definitive, vertical conduit of commerce, next to the horizontal one that has nourished it for centuries.

At the bottom of the tower, the connected building is provided with photovoltaic roofing and thermal barrier coatings, which serve to insulate components from large and prolonged heat loads by utilizing thermally insulating materials, which can sustain an appreciable temperature difference between the load-bearing alloys and the coating surface.

Moreover, on the public square, large sunshade screens made of perforated black panels provide shelter for pedestrians and also act as wind breakers. Some of these umbrellas on the lower square are planned to become supports for plantings, in the second development phase, supported by an automatic watering system.

6 Bevis Marks

London, United Kingdom

A new, highly efficient building designed to provide Grade-A office space and retail in the heart of the City, 6 Bevis Marks replaces an outdated 1980s structure. The redesign retained the existing basement retaining wall and piled foundations. The new building is nine floors taller than its predecessor and includes three private roof terraces, a sky court on the roof, and more than double the previous public space on the site. On the lower floors the elevations are a modern interpretation of the richly textured cladding of the nearby Holland House. This is most apparent where the cladding is brought forward of the glass façade line to read almost as a solid mass when viewed obliquely.

An ETFE floating roof structure turns the rooftop into an all-weather, open-air, landscaped space with spectacular views. The roof structure drops in front of the four uppers floors to help support the cantilevered shading. Despite doubling the floor area of the original building, the new structure is 80 percent more energy-efficient.

Completion Date: June 2014
Height: 74 m (245 ft)
Stories: 17
Area: 20,700 sq m (222,813 sq ft)
Use: Office
Owner: AXA; BlackRock; Wells Fargo
Developer: CORE
Architect: Fletcher Priest Architects
Structural Engineer: Waterman Building Services
MEP Engineer: Waterman Building Services
Main Contractor: Skanska
Other Consultants: Applied Acoustics (acoustics); NET Project Management and Consultancy Services (façade); Ramboll (fire); Townshend Landscape Architects (landscape); WT Partnership (quantity surveyor)

Opposite: Overall view of tower from east

Top: View of ETFE floating roof structure and rooftop garden overlooking skyline

Above: Main lobby

Right: Typical section

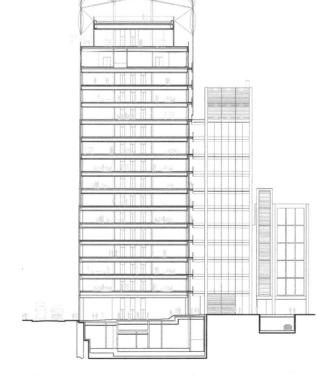

10 Brock Street
London, United Kingdom

The latest addition to the 13-acre mixed-use campus at Regent's Place, 10 Brock Street marks the final phase of the 17-year redevelopment of this central London estate. The distinctive building is arranged as three slender towers, running front-to-back, following the principles of the permeable master plan. The staggered sizes and angled planes of the volumes also help to mediate between the height of the adjacent Euston Tower and the adjoining buildings to the west, suggesting a spiraling movement in composition.

The leitmotif of the three towers, with their angled tops and plans, has been used to inform the detailing of the façades and interiors. The atrium design has enabled the floor plates facing south across the plaza to the main street to be opened up across the full width of the building, and also created a large space for the soaring entrance lobby, which acts as a continuation of the large public square in front of the building.

Completion Date: July 2013
Height: 72 m (236 ft)
Stories: 16
Area: 49,239 sq m (530,004 sq ft)
Use: Office
Owner/Developer: British Land
Architect: Wilkinson Eyre Architects
Structural Engineer: CH2M Hill
MEP Engineer: Watkins Payne Partnership
Project Manager: M3 Consulting
Main Contractor: Lend Lease
Other Consultants: AECOM / Davis Langdon (cost); Arup (façade); Maurice Brill Lighting Design (lighting)

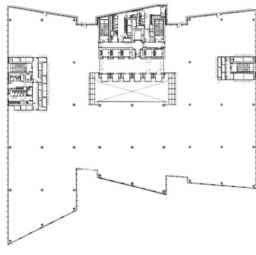

Opposite Top: Overall view of building from southeast

Opposite Bottom: Overall view from the North

Top Left: Typical section

Bottom Left: Typical floor plan

Top Right: View of entrance and its relation to plaza

Bottom Right: Main entrance and lobby

AvB Tower
The Hague, Netherlands

The AvB Tower is a "hyper-hybrid" academic building and a dynamic hub, threading The Hague's central train station to the adjacent urban envelope, and accommodating students and faculty with a program of housing, retail, and dining. Surrounded by an expansive and newly constructed pedestrian square – the Anna van Buerenplein – this steel tower is foreseen as a catalyst for further redevelopment of this centrally sited urban area. The square is a point of convergence for multiple modes of transportation and a gathering space for the tower's 396 student residents.

Clusters of office buildings and a surface parking lot were demolished to create this new residential and retail project, which is part of a larger master plan to restructure an under-utilized space in the center of the city. A newly submerged three-story parking garage serves as the plinth for the tower's foundation. The pre-existing weight limitations of the parking garage dictated the use of structural steel, distributing its load across 11 pre-determined columns.

Completion Date: September 2013
Height: 71 m (233 ft)
Stories: 22
Area: 25,500 m (274,280 sq ft)
Use: Residential/Educational
Owner: Green College Court BV
Developer: Anna van Buerenplein BV
Architect: Wiel Arets Architects
Structural Engineer: DGMR Raadgevende Ingenieurs
MEP Engineer: Van Rossum Raadgevende Ingenieurs

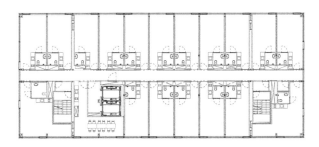

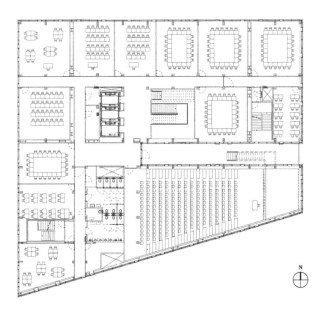

N

Opposite: Overall view from northwest

Left: Floor plans – typical student housing level (top) and typical classroom level (bottom)

Top Right: Sculptural staircase designed to encourage informal social exchanges

Bottom Right: Overall view from southeast

CalypSO
Rotterdam, Netherlands

The CalypSO building is a component of the Rotterdam Central urban strategy scheme. The development of this site afforded the opportunity to establish the beginnings of a quality public realm on the Westersingel leading from the station, the desired "cultural route" of the city. The development accommodates 407 apartments, a retail area, commercial space, and a church. The faceted group of towers, with the more rotund, copper-clad Pauluskerk church nestling against it, is like a collection of crystalline rocks. Punched into which are sheltered external spaces, which are generous enough to be used properly, and integrated into the internal planning so as to form an extension of the living spaces.

The quality of the public areas in particular have been considered, with a two-tiered circulation space below the residential buildings; the lower below the transparent pavement of the upper, and connecting directly with car parking. Consultation and workshop exercises with the community of Pauluskerk in particular, were central to the development of the design.

Completion Date: March 2013
Height: Tower 1 & 2: 71 m (234 ft)
Stories: Tower 1: 22; Tower 2: 23
Use: Tower 1: Residential/Office; Tower 2: Residential
Owner/Developer: De Wilgen Vasrgoed
Architect: Alsop Architects; Van Der Laan Bouma Architekten BV
Structural Engineer: Van Der Laan Bouma Architekten BV
MEP Engineer: Van Der Laan Bouma Architekten BV
Main Contractor: Boele & van Eesteren
Other Consultants: Cauberg-Huygen Consulting Engineers (acoustics); dS+V (planning)

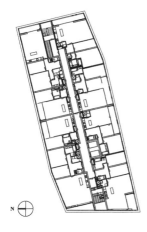

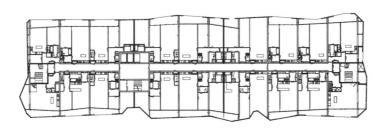

N

Opposite Top: Overall view of towers from southwest

Opposite Bottom: Overall view of towers from northwest

Top: Typical floor plan

Left: View of Lobby, showcasing how the façade impacts the interior space

Top Right: Façade detail

Bottom Right: View of church nestled against the towers, illuminated from within at night

Exzenterhaus Bochum
Bochum, Germany

The Exzenterhaus is a major orientation point for the city of Bochum, Germany. Its shape, with 15 polycentric rings set atop the solid foundation of a 22-meter-tall bomb shelter from the World War II era, would probably make it a landmark at any height. Each floor contains rounded, rentable office space, offering 360-degree views and copious natural light through full-height windows and balconies up to 1.6 meters deep. Communicating spiral staircases between floors in single-tenant spaces are also an option.

The bunker shaft provides a two-story reception area at the base. The remaining bunker floors serve as archival and storage space. Further uses may include light displays and use as a projection screen on the exterior. The structural design of the building is a tour-de-force. The concrete structure's floor slabs cantilever up to 4.5 meters but are only 250 millimeters thick. This is supported by a conical spatial form activating membrane action in the slabs, as well as unbounded, pre-stressed tendons around the slab edges, to counteract cracking.

Completion Date: August 2013
Height: 89 m (291 ft)
Stories: 25
Area: 10,400 sq m (111,945 sq ft)
Use: Office
Owner: Exzenterhaus Bochum GmbH & Co. KG
Architect: Gerhard Spangenberg Architekt
Structural Engineer: Schlaich Bergermann und Partner; GuD GmbH; Drees & Sommer Advanced Building Technologies
MEP Engineer: HHP Nord; Ing. Büro Landwehr GmbH; Kleinmann Engineering GmbH
Project Manager: Arte Baumanagement GmbH

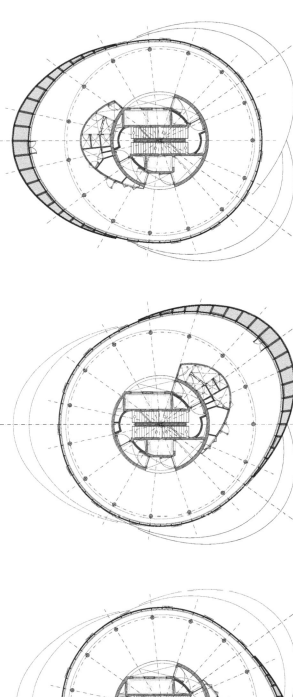

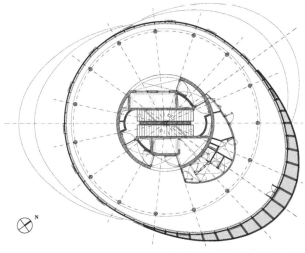

Opposite: Overall view of tower from south

Top Left: Main Entrance

Bottom Left: Detail view looking up

Right: Typical floor plans – levels 16 to 20 (top), levels 11 to 15 (middle), and levels 6 to 10 (bottom)

Fletcher Hotel Amsterdam
Amsterdam, Netherlands

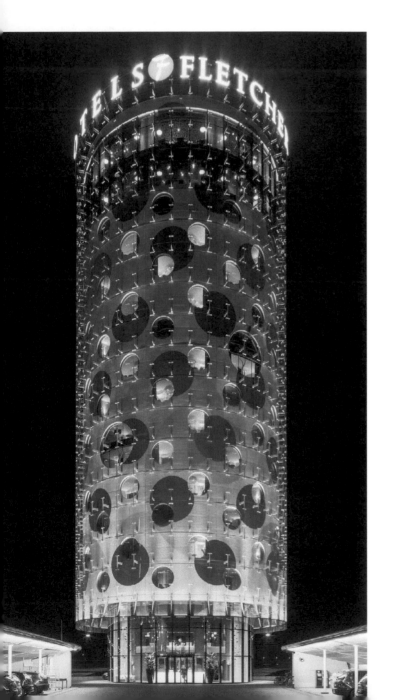

The Fletcher Hotel is a new landmark for Amsterdam. The 60-meter tower has a compact, circular floor plan with a diameter of merely 24 meters, resulting in a characteristic slim silhouette alongside the highway. The objective was to create an omnidirectional structure, with an expressive façade. The resulting limited space is used as efficiently as possible. Service areas and technical spaces are situated in the basement, in the pedestal or on the roof.

Guest rooms encircle the staircase and lifts in the heart of the hotel. On the 15th level, five board rooms have been arranged in a manner that allows them to be linked together, while sky lounges and restaurants offer panoramic views from the two floors above. The fully glazed façade, with its bending screens and round windows, lends a distinctive and yet restrained transparent appearance to the building. The construction challenge of hanging a glass wall on a lightweight inner façade was solved by using glass-reinforced plastic composite elements.

Completion Date: January 2013
Height: 60 m (197 ft)
Stories: 17
Area: 7,000 sq m (75,347 sq ft)
Use: Hotel
Owner: Fletcher Hotel Group
Developer: M. Caransa BV
Architect: Benthem Crouwel Architects
Structural Engineer: Van Rossum Raadgevende Ingenieurs
MEP Engineer: Wichers & Dreef
Project Manager: M. Caransa BV
Main Contractor: Strukton Bouw
Other Consultants: Kolenik Eco Chic Design (interior); Living Design (lighting); moBius consult (building physics); Octatube (façade)

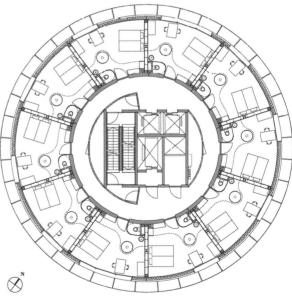

Opposite: Overall view of tower from east

Top Left: Detail view of façade from below

Bottom Left: Overall view of building, approaching from south on the highway

Top Right: Typical hotel room

Bottom Right: Typical floor plan

Maslak Spine Tower
Istanbul, Turkey

This mixed-use tower project embodies a circular spatial configuration, instead an angular bearing, avoiding overwhelming its surroundings. Yet it is likely to capture attention from multiple angles. Its distinctive, iconic presence directly reflects the aspirations of the business environment of Istanbul, which is gathering strength in global terms. The success of the design comes from efforts to alleviate and decompose the building's mass by way of several slight but attentive touches along its exterior.

These include an ovoid cut in the façade, which breaks up the monolithic appearance of the glass curtain wall and reminds the onlooker of its curving form in plan even when viewed straight on from the side. On the reverse side of the tower, the skin is pulled outwards where the cut appears, as if a second parabolic form were slowly passing through the tower. The rounding continues in a third plane, scalloping the top eight floors under a curving glass atrium.

Completion Date: September 2014
Height: 202 m (663 ft)
Stories: 47
Area: 137,500 sq m (1,480,038 sq ft)
Use: Residential/Office
Owner: Soma Group
Developer: Tilaga
Architect: iki design group
Structural Engineer: Erdemli Engineering
MEP Engineer: Atakar Engineering; SMT Engineering
Main Contractor: Tilaga
Other Consultants: Elsan Consultancy (energy concept); Etik Consultancy (fire); Harr Holand Consultancy (lighting); Japsen Consultancy (vertical transportation); Metal und Fassaden Consultancy (façade); Ruscheway Consultancy (wind)

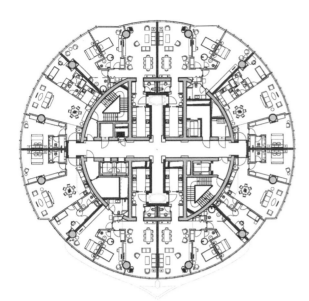

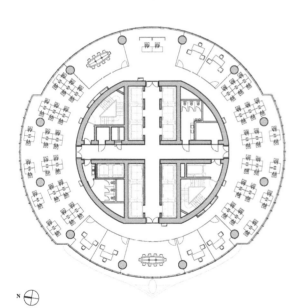

N ⊕

Opposite: Overall view of top half of tower

Above: Typical floor plans – typical residential (top) and typical office (bottom)

Top Right: Tower crown

Bottom Right: Overall night view of building in context

Solaria
Milan, Italy

Solaria gives its residents an exclusive view, from the spires of the Duomo to the Alps. The building's external architecture is composed of three separate wings that meet in a central core, from which natural light radiates into every floor. Large, transparent, light-filled volumes and windows up to seven meters tall are emphasized by furnishings created expressly for this building. The elegance of the main lobby is a contemporary reinterpretation of the prestige of classical residential architecture.

Solaria is strategically located in Milan. It has been carefully designed to meet residents' demands for privacy while, at the same time, providing them with the convenient links to the urban fabric that they need. There are three metro lines within walking distance. Moreover, Solaria overlooks the Giardini di Porta Nuova park to the north, and to the south, a courtyard garden of 4,000 square meters, linked to Piazza della Repubblica by a pedestrian path.

Completion Date: November 2013
Height: 143 m (469 ft)
Stories: 37
Area: 25,000 sq m (269,098 sq ft)
Use: Residential
Owner/Developer: Hines Italia
Architect: Arquitectonica
Structural Engineer: Arup
MEP Engineer: Deerns; Ariatta Ingegneria
Project Manager: Jacobs Engineering
Main Contractor: CMB
Other Consultants: Coima Image (interiors); Galotti Spa (construction management); J&A (quantity surveyor); Kohn Pedersen Fox Associates (urban planner); LAND (landscape)

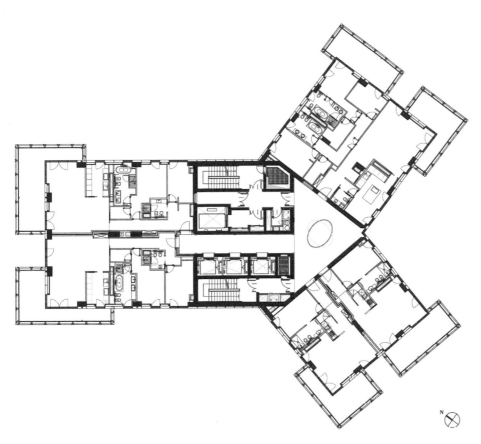

N

Opposite: Overall view of tower from south

Top: Typical floor plan

Left: Typical interior space

Right: Detailed view of balconies

The Tower, One St George Wharf

London, United Kingdom

Located on a prominent site on a sharp bend of the River Thames, the Tower is a cylindrical building providing 211 luxury apartments over 50 stories. At just over 180 meters, it is one of Europe's tallest wholly residential buildings and the tallest all-residential tower in London. The unique floor plan concept is typically divided into five apartments per floor, with separating walls radiating out from the central core. Sky gardens provide residents with a semi-external space, stepped forward from the pure circular plan; with floor-to-ceiling glazing providing great 360-degree views across London.

Despite its glassy appearance, the building uses significantly less energy than that of a conventional tall building. A sophisticated, ventilated cavity façade, with operable windows to reduce reliance on air conditioning, achieves significant CO_2 emission savings over the building's lifespan. Renewable energy sources are also utilized. A 10-meter-high vertical axis wind turbine crowns the structure, generating an expected 27,000 kWh of energy per year, powering lighting in communal areas.

Completion Date: March 2014
Height: 181 m (594 ft)
Stories: 52
Area: 30,342 sq m (326,599 sq ft)
Use: Residential
Owner/ Developer: St George South London
Architect: Broadway Malyan
Structural Engineer: White Young Green; Robert Bird Group
MEP Engineer: Grontmij
Main Contractor: Brookfield Multiplex; St George South London
Other Consultants: AECOM / Davis Langdon (cost); Barton Wilmore (planning); BMT Fluid Mechanics, Ltd. (wind); Broadway Malyan (landscape); Cladtech Associates (façade); Cole Jarman (acoustics); Grontmij (fire); REEF Associates Ltd (façade access)

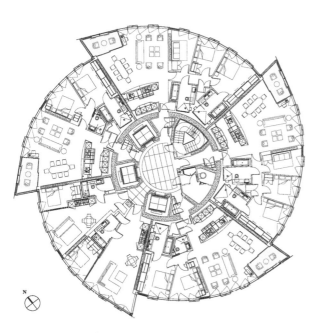

Opposite: Overall view of tower from southwest

Top Left: Typical residential floor plan

Bottom Left: Interior living space

Top Right: Enclosed "sky garden" balcony, location shown in shaded portion of the floor plan

Bottom Right: Building entrance

Tour Carpe Diem
Paris, France

Tour Carpe Diem sets La Défense on a fresh path to pedestrian-friendly urbanism by reconnecting the raised esplanade that continues the axis of the Champs-Elysées to the surrounding urban fabric of the town of Courbevoie to the north. A monumental stair descends from the building's winter garden and lobby to a public plaza on the Boulevard Circulaire, where a second front welcomes visitors at what was previously the back of the site.

The building's pure form and faceted façades allow the tower to stand out in a thicket of tall buildings in La Défense. The building's sleek glass façades are stretched tautly over distinctive sculptural crystalline massing generated by careful consideration of climate and context. Sweeping window walls are oriented to the best views, and narrow façades are turned to the setting sun to minimize solar gain in the office interiors. In addition, the tower is topped with a spacious roof garden, and has achieved LEED-CS Platinum certification.

Completion Date: September 2013
Height: 166 m (545 ft)
Stories: 35
Area: 45,500 sq m (473,784 ft)
Use: Office
Owner: Aviva France; Crédit Agricole Assurances
Architect: Robert A.M. Stern Architects (design); SRA Architectes (architect of record)
Structural Engineer: Terrell Group
MEP Engineer: SNC Lavalin
Project Manager: EPADESA
Main Contractor: SPIE Batignolles; Besix
Other Consultants: Atelier Ten (sustainability); Cabinet Daniel Legrand (quantity surveyor); Cabinet Penicaud (now part of SNC Lavalin) (sustainability); GV Ingenierie (cost); Hines France (development manager); Jean Jegou (interior); Lasa (acoustics); Mutabilis (landscape)

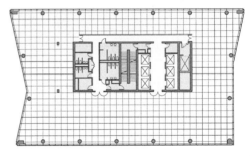

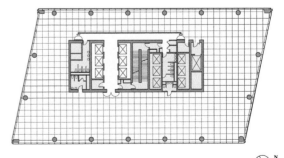

Opposite: Overall view of tower from northeast

Top Left: View of the penthouse roof garden

Bottom Left: Building entrance with monumental stair

Top Right: Interior winter garden

Bottom Right: Floor plans – 24th level (top) and 7th level (bottom)

E' Tower

Eindhoven, Netherlands

Completion Date: October 2013
Height: 68.5 m (236 ft)
Stories: 20
Area: 16,000 sq m (172,223 sq ft)
Use: Residential
Owner: Stadionkwartier BV
Developer: Stadionkwartier BV
Architect: Wiel Arets Architects
Structural Engineer: Nelissen BV
MEP Engineer: Tielemans BV
Main Contractor: Scheldebouw BV

Grand Office

Vilnius, Lithuania

Completion Date: May 2014
Height: 82 m (270 ft)
Stories: 22
Area: 14,010 sq m (150,802 sq ft)
Use: Office
Owner: YIT
Developer: YIT
Architect: Architektūros Linija; Viltekta
Structural Engineer: Viltekta
MEP Engineer: Viltekta
Main Contractor: YIT

One Angel Square

Manchester, United Kingdom

Completion Date: January 2013
Height: 67.5 m (221 ft)
Stories: 14
Area: 47,500 sq m (511,286 sq ft)
Use: Office
Owner: The Co-operative Group
Developer: The Co-operative Group
Architect: 3DReid
Structural Engineer: Buro Happold
MEP Engineer: Buro Happold
Main Contractor: BAM Construction
Other Consultants: Gardiner & Theobald Inc (cost)

Solea

Milan, Italy

Completion Date: November 2013
Height: 72 m (235 ft)
Stories: 15
Use: Residential
Owner/Developer: Hines Italia
Architect: Caputo Partnership
Structural Engineer: Arup
MEP Engineer: Deerns; Ariatta Ingegneria
Project Manager: Jacobs Engineering
Main Contractor: CMB
Other Consultants: Coima Image (interiors); Galotti Spa (construction management); J&A (quantity surveyor); Kohn Pedersen Fox Associates (urban planner); LAND (landscape)

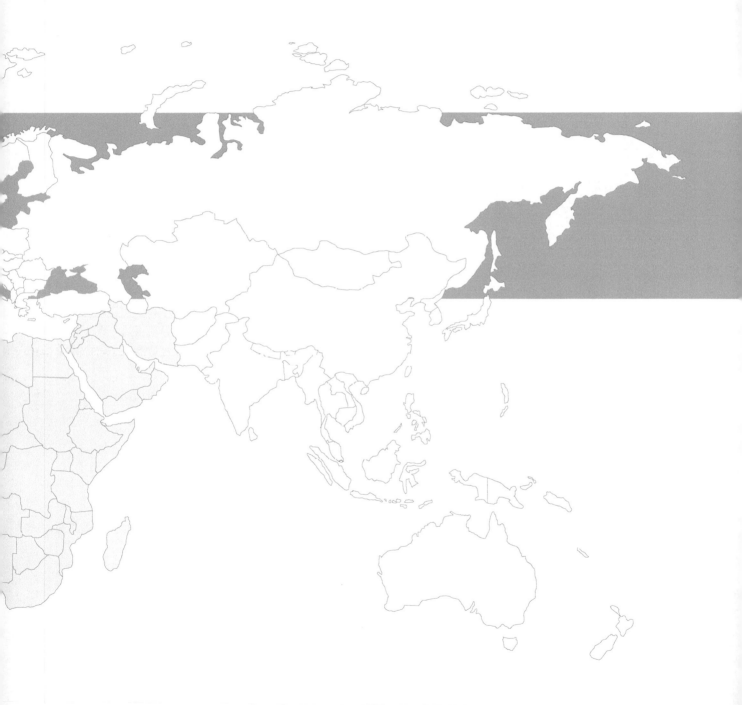

Best Tall Building
Middle East & Africa

Cayan Tower
Dubai, United Arab Emirates

"The building presents itself differently from every angle through its dynamic form, making it an exceptional eye catcher in the skyline of Dubai."

David Gianotten, Juror, OMA

The Cayan Tower is a 75-story luxury apartment building with a striking helical shape, turning 90 degrees over the course of its 306-meter height. Each floor plate is identical, but is set 1.2 degrees clockwise from the floor below, giving the tower a distinctive form by way of an innovative, efficient, repeatable structure. Its shape is a pure expression of the relationship between a building's form and the structural framework that supports it.

Aesthetically, the twisting shape makes the building stand out from the architectural disharmony of the Dubai waterfront, which is largely composed of indistinct towers that do not speak to their location. Located near Dubai Internet City, Emirates Golf Club, and numerous corporate headquarters, the tower's twisting form provides a greater number of units with desirable views of the Dubai Marina and Arabian Gulf, while also preserving the views for residents living in neighboring buildings, ensuring that Cayan Tower enhances its spectacular waterfront site.

Completion Date: June 2013
Height: 306 m (1,005 ft)
Stories: 73
Area: 111,000 sq m (1,194,794 sq ft)
Use: Residential
Owner/Developer: Cayan Investment & Development
Architect: Skidmore Owings & Merrill (design); Khatib & Alami (architect of record)
Structural Engineer: Skidmore, Owings & Merrill (design); Khatib & Alami (engineer of record)
MEP Engineer: Skidmore Owings & Merrill
Project Manager: Currie & Brown
Main Contractor: Arabtec
Other Consultants: Alan G. Davenport Wind Engineering Group BLWTL (wind); Cerami Associates (acoustics); Fisher Marantz Stone (lighting); Lerch Bates (vertical transportation); Opening Solutions, Inc. (vertical transportation); Rolf Jensen & Associates (fire); Sako & Associates, Inc. (security); Shen Milsom Wilke, Inc. (acoustics); SWA Group (landscape); Van Deusen & Associates (vertical transportation)

The tower's structural system is a cast-in-place, high-strength, reinforced-concrete column superstructure. The shape and size of the columns were determined through the use of wind-tunnel testing and three-dimensional computer modeling to analyze building stresses. The building core is a cylindrical concrete form that acts as the central pillar for the tower. As the building ascends, the rotation at each floor occurs around this central mass. The building was constructed using a "jump form" system that takes advantage of its repetitive nature.

Within the tower, 495 units have been divided into six unique types, offering ample living space configurations to residents. Standard units range from studios to three-bedrooms, and the crown of the building features six levels of both half- and full-floor penthouses. Residential amenities include an outdoor infinity edge pool facing the marina, flexible spaces on the sixth floor, a health spa, exercise facility, and conference center.

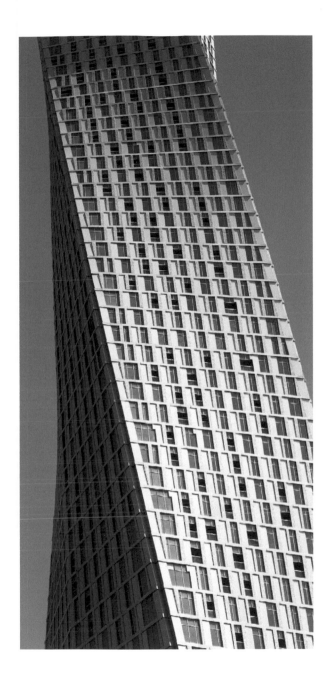

Previous Spread

Left: Overall view of tower

Right: Close up view of façade

Current Spread

Opposite Top: View of the tower in the Dubai Marina context

Opposite Middle: Interior space showing the sun-shading metal screens

Opposite Bottom: Curving hallway corridor

Left: Close up view of the twisting tower

On the ground plane, the tower acts as an urban gateway, connecting the Arabian Gulf, the Dubai Marina, and the city beyond. It also provides a visual point of reference on the city skyline. The building appears to change from every angle, giving it a sense of movement regardless of the viewer's vantage point. Within the marina, the public can experience the design detail firsthand by walking along the waterside promenade that edges the tower's site. A retail colonnade, located at the tower's base, also provides visitors with shopping, shade, and views out toward the water.

The tower's twisting shape is designed for enhanced indoor comfort. Its twist generates self-shading for the tower, ensuring that many of the interior spaces are protected from solar exposure. The building's exterior terraces and the façade's metal cladding panels, high-performance glass, and deep sills around the recessed glass line further protect the building from direct solar radiation, while providing diffuse daylight to

"The intelligent helical design of the Cayan Tower responds to very specific and challenging local conditions, while providing a visually striking new landmark for the Dubai skyline. This building expresses its structure through its form in an elegant and sophisticated way, enhancing the architecture of the existing waterfront site."

Sir Terry Farrell, Juror, Farrells

Jury Statement

The twisting form of the Cayan Tower is an unusually elegant statement. The juxtaposition of the 75-story tower next to the water, and next to the nearby rectangular buildings helps to create a softer, inspirational and exciting urban environment that would not have been possible with a conventional tower. The engineering is exceptional, and it is praiseworthy not only for its visual impact, but also for its material economy and appropriate responsiveness to a challenging urban and environmental condition. The attention to detail, including the subtly angled perforated titanium screens that moderate the harsh desert light, transcends scale and is impressive from both inside and out.

In an environment where so many tall buildings lined up in a row against a humid and reflective backdrop can make massive buildings seem like cardboard cut-outs, it takes an extraordinary design gesture to indelibly express the three-dimensionality of a building. Cayan Tower makes that gesture; happening upon its dancing form in the skyline is like encountering a hula-hooper on a train full of grey flannel suits.

interior spaces. This enhanced design for solar control reduces the building's demand for cooling, provides a thermally comfortable environment, and minimizes the risk of glare, while optimizing occupant views of the surrounding marina environment and gulf.

The tower's HVAC system has been specially designed to deal with desert conditions. A central dedicated outside air system equipped with sand filters and heat pipes distributes fresh air across the tower. Fan coil units within the occupied space satisfy the cooling load while providing additional local filtering to reduce the level of fine particles entering through the façade.

The tower's helical form also acts as a shield from the northerly diurnal winds, which often carry sand and dust, thus minimizing the fine particles that may flow through the façade and impact indoor air quality. When outdoor conditions allow, windows can be opened so that natural ventilation can provide fresh air and passive cooling in interior spaces. At night, when cool winds blow from east and west, a separate system that passively cools the tower's slabs helps the building discharge excess heat. Outside air is naturally brought into the space and used to "purge" heat absorbed by the tower's exposed slabs during the warmer day hours. This cools down the thermal mass and restores the slabs' thermal properties, enabling them to again absorb heat during the following day.

Left: Overall view at dusk in the context of other Dubai Marina skyscrapers

Opposite Left: Typical section

Opposite Right: Floor plans – unit type five, penthouse layout (top), and unit type three layout (bottom)

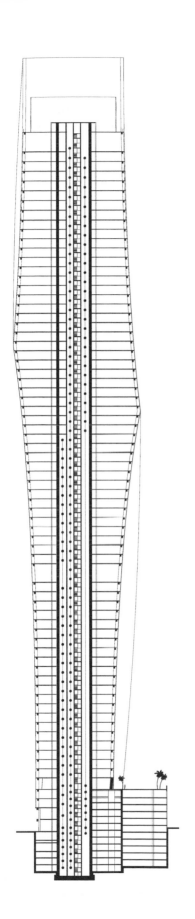

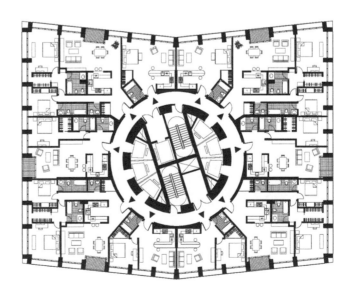

BSR 3
Tel Aviv, Israel

The BSR 3 business center is located at the intersection of three of the busiest commercial and industrial districts of the Tel Aviv metropolitan area. Each office floor plate spans 8.2 meters, column-free from the core to the exterior, allowing maximum flexibility for space planning.

The design addresses the region's local architecture by taking the residential balcony and roof terrace, spaces that take advantage of the year-round moderate climate, and translating this into a business environment. The simple, but not simplistic, approach was to "move" eight to ten floors at the upper levels of the tower in a staggered layout, so that large shaded balconies were created. An array of red terra-cotta "boxes" penetrate the tower's skin at strategic points, creating more unique office spaces. The result is a bold and well-defined vertical building with an interesting overall look.

Completion Date: March 2013
Height: 145 m (476 ft)
Stories: 36
Area: 41,393 sq m (445,551 sq ft)
Use: Office
Owner/Developer: B.S.R. Group
Architect: Y.A. Yashar Architects, Ltd.
Structural Engineer: David Engineers, Ltd.
MEP Engineer: HRVAC Consulting Engineering Co., Ltd.; Krashin Electrical Engineering Consulting, Ltd.; A. Papish & Co. Consulting Engineering, Ltd.
Main Contractor: Ortam – Malibu Group
Other Consultants: Aldaag Engineers Consultants, Ltd. (fire); El-Rom Consulting Engineer, Ltd. (vertical transportation); Susana Lihu Landscape Architecture (landscape); Dagesh Engineering Traffic & Road Design, Ltd. (traffic)

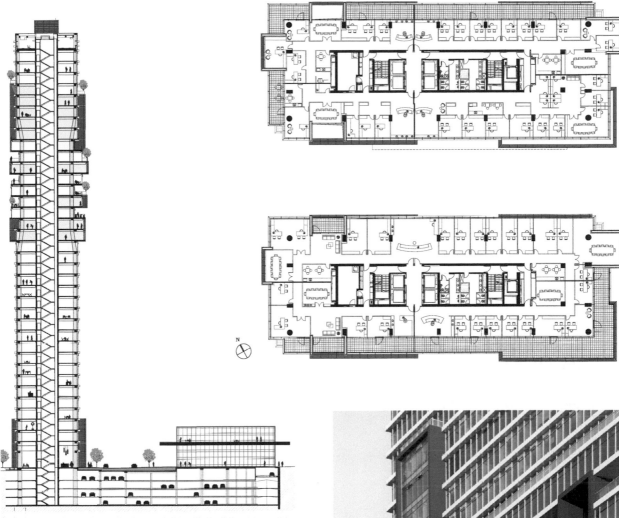

Opposite: Overall view of tower from southwest

Above: Typical section

Top Right: Floor plans – level 24 (top), and level 23 (bottom)

Bottom Right: Detail view of façade variation and shaded balconies

Champion Tower

Tel Aviv, Israel

The design of the Champion Tower is a reconciliation between the commercial and office spaces, and between two different objectives for the commercial space at ground level. The main challenge was to connect the two elements, while retaining and enhancing the street fabric. The project owner wanted to accommodate two distinctive car showrooms for Volkswagen and Audi, which led to the idea of breaking the horizontal element in order to create a more human-scaled street frontage for the development. The length of the development was also broken down by way of a pedestrian passageway through the base of the building.

The tower was similarly divided into two distinct vertical elements that have a dialogue with each other. The separation is emphasized by the recessed balconies and the change in cladding between the elements. The dialogue between elements extends up the tower, and from one side to the other, and is tempered by the grey horizontal lines binding the two sides together.

Completion Date: March 2014
Height: 165 m (541 ft)
Stories: 42
Area: 65,785 sq m (708,104 sq ft)
Use: Office
Owner/Developer: Allied Real Estate, Ltd.; Migdal group
Architect: MYS Architects
Structural Engineer: David Engineers, Ltd.
MEP Engineer: Doron Shachar Engineers; Bar Akiva Engineers; Ada Bronfman Engineers & Consultants Ltd
Main Contractor: Omer Construction & Engineering, Ltd.
Other Consultants: Dagesh Engineering Traffic & Road Design, Ltd. (traffic); Feder Architects (access); Kobi Gamzo (cost); Leshem Sheffer Environmental Quality, Ltd. (environmental); Liora Niv Fromchenko (interior); M.G Acoustic Consultants, Ltd. (acoustics); S. Lustig Engineers & Consultants, Ltd. (vertical transportation); RTLD Lighting Design (lighting); Zur Wolf Landscape Architects (landscape)

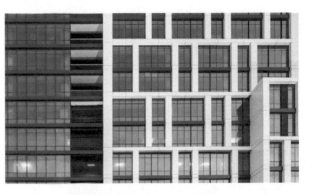

Opposite: Overall view of tower from northeast

Above: Tower at sunset

Top Right: VW showroom and relation to the street

Middle Right: Façade detail

Bottom Right: View of lobby

Portside
Cape Town, South Africa

Portside is the first high-rise to ascend in Cape Town for more than two decades, and is now its tallest. The banking headquarters building is becoming a landmark in the emerging financial district of Cape Town.

The building is about 10 meters shorter than the permitted height, so as to limit visual impact in relation to the larger urban and geographical context of the city and Table Mountain. The architectural quality and material selection were crucial in terms of mediating Portside's relationship to the public realm. Street level activation, appropriate scale, legibility and permeability were the main vehicles for addressing social context and urban regeneration. The use of LED lighting, storm water recapture, low-embodied-energy materials, and the provision of individually marked, removable and recyclable unitized façade panels earned Portside a 5 Green Star Design rating, the first tall building in South Africa to do so.

Completion Date: April 2014
Height: 139 m (456 ft)
Stories: 30
Area: 114,547 sq m (1,232,974 sq ft)
Use: Office
Owner: Old Mutual Properties; First Rand Bank
Developer: Old Mutual Properties; Eris Property Group
Architect: dhk architects; Louis Karol Architects
Structural Engineer: WSP Group; Aurecon (peer review)
MEP Engineer: Spoormaker and Partners; Claassen Auret International; BFBA; Benatar Consulting; Matrix Consulting Services (peer review)
Project Manager: SIP Project Managers; Metrum Project Management; Absolute Project Management
Main Contractor: Murray & Roberts
Other Consultants: AECOM (quantity surveyor); Agama Energy (environmental); Clive Newsome (civil); De Leeuw group (quantity surveyor); MacKenzie Hoy Consulting Acoustic Engineers (acoustics); Proji-tech (vertical transportation) Solution Station (fire); WAC Projects (vertical transportation)

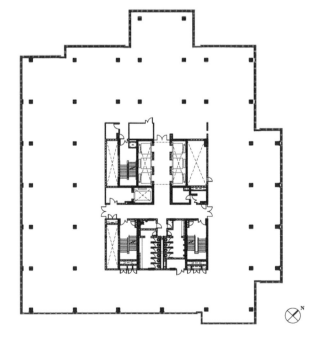

Opposite: Overall view of tower from east

Top: Tower with port and skyline context

Left: View of main entrance and lobby

Right Top: Façade detail looking up

Right Bottom: Typical floor plan

The Landmark
Abu Dhabi, United Arab Emirates

The Landmark is a 72-story office and residential tower on the Corniche, Abu Dhabi's grand waterfront crescent. Standing apart from the city's other tall buildings, the tower is visible on all sides and has panoramic views of the Persian Gulf and the surrounding islands. The Landmark's design – the winner of an international competition – uses local precedents to be environmentally sustainable and culturally sensitive.

To address the challenging desert weather conditions, the Landmark was conceived as a series of layered screens, which form a protective wrapper extending from the building's conditioned envelope. While the building is contemporary in appearance, this approach evokes the use of screens in vernacular Arabic architecture. The plan of the building also has a cultural precedent. Its geometry is based on the dodecagon, the 12-sided figure frequently used in Islamic art. The tower top, which hosts a substantial sky garden, uses the temperature gradient and higher wind speed to reduce the need for cooling, a traditional practice in the Gulf region.

Completion Date: June 2013
Height: 324 m (1,063 ft)
Stories: 72
Area: 158,645 sq m (1,707,641 sq ft)
Use: Residential/Office
Architect: Pelli Clarke Pelli Architects
Structural Engineer: Buro Happold
MEP Engineer: Buro Happold
Main Contractor: Al Habtoor Engineering; Consolidated Contractors International Company
Other Consultants: AECOM / Davis Langdon (quantity surveyor); Applied Landscape Design (landscape); Alan G. Davenport Wind Engineering Group BLWTL (wind); EPPAG (façade); Hilson Moran (vertical transportation); Isometrix (lighting); REEF Associates Ltd (façade maintenance)

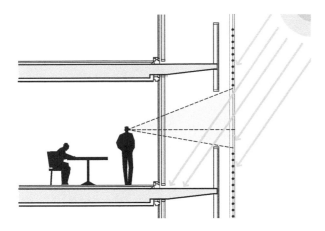

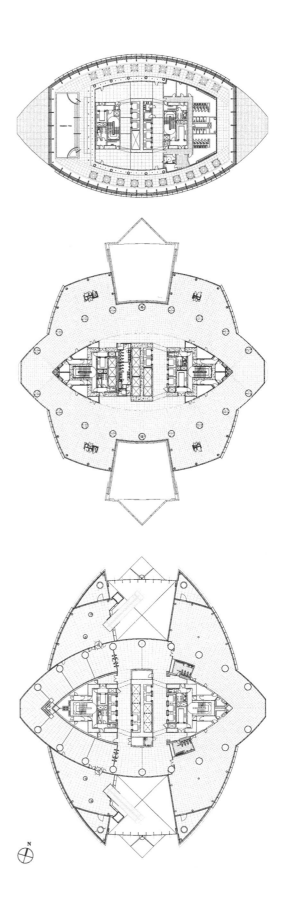

Opposite: Overall view of tower from south

Top Left: Section drawing through façade, showing shading screen feature

Bottom Left: Portion of the sky garden located at the top of the tower

Right: Floor plans – level 67 sky garden (top), level 4 (middle) and second floor lobby (bottom)

22 Rothschild Tower
Tel Aviv, Israel

Completion Date: February 2014
Height: 127 m (416 ft)
Stories: 29
Area: 32,175 sq m (346,329 sq ft)
Use: Hotel/Office
Owner: Aviv & Co., Ltd.; LR Group, Ltd.
Developer: Aviv & Co., Ltd.
Architect: Moshe Tzur Architects and Town Planners, Ltd.
Structural Engineer: Rami Ballas Engineering, Ltd.
MEP Engineer: E. Lynn – S. Cohen Plumbing Consultant, Ltd.; M. Doron – I. Shahar & Co. Consulting Eng., Ltd.; Sahi Harpaz – Electrical Consulting & Eng., Ltd.
Main Contractor: Aviv & Co., Ltd.
Other Consultants: Dagesh Engineering Traffic & Road Design, Ltd. (traffic); Eng. S. Lustig – Consulting Engineers, Ltd. (vertical transportation); K.A.M.N. Structural Protection Consulting (civil); S. Mashiah Consultants in Acoustics, Ltd. (acoustics)

Conrad Hotel
Dubai, United Arab Emirates

Completion Date: September 2013
Height: 251 m (823 ft)
Stories: 51
Area: 134,729 sq m (1,450,210 sq ft)
Use: Hotel/Office
Owner/Developer: Private Property Management, Abu Dhabi
Architect: Smallwood, Reynolds, Stewart, Stewart (design); Atkins (architect of record)
Structural Engineer: BECA Group; Atkins
MEP Engineer: BECA International, Ltd.; Atkins
Main Contractor: Arabtec; Dubai Contracting Company
Project Manager: International Project Management
Other Consultants: Bo Steiber Lighting Design (lighting); Fortune Consultants, Ltd. (vertical transportation); Meinhardt (façade); Peridian Asia Pte., Ltd. (landscape); Windtech Consultants (wind)

Rosewood Abu Dhabi
Abu Dhabi, United Arab Emirates

Completion Date: May 2013
Height: 140 m (459 ft)
Stories: 36
Area: 102,100 sq m (1,098,995 sq ft)
Use: Residential/Hotel
Owner/Developer: Mubadala Development Company
Architect: Handel Architects
Structural Engineer: Magnusson Klemencic Associates
MEP Engineer: WSP Group
Project Manager: EC Harris
Main Contractor: Arabian Construction
Other Consultants: AECOM (landscape); Brennan Beer Gorman Monk (interiors); Cline Bettridge Bernstein Lighting Design, Inc. (lighting); EC Harris (cost); Haley & Aldrich (geotechnical); Meinhardt (façade); Shen Milsom Wilke, Inc. (acoustics); RFR (façade); Viridian Energy & Environmental, LLC (environmental); WSP Group (civil)

World Trade Center Doha
Doha, Qatar

Completion Date: March 2014
Height: 242 m (794 ft)
Stories: 51
Area: 141,968 sq m (1,528,131 sq ft)
Use: Office
Owner/Developer: Qatar General Insurance & Reinsurance, Co.
Architect: MZ & Partners
Structural Engineer: MZ & Partners
MEP Engineer: MZ & Partners
Project Manager: Projacs International
Main Contractor: Arabtec
Other Consultants: BMT Fluid Mechanics, Ltd. (wind)

CTBUH Urban Habitat Award

Award Criteria

Recognizing that the impact of a tall building is far wider than just the building itself, the CTBUH has introduced a new "Urban Habitat" award in 2014. This award is intended to recognize significant contributions to the urban realm, in connection with tall buildings. Submissions could thus range from a brilliantly executed master plan which has led to a quality urban environment, or at the scale of a single site, where the interface between tall building and the urban realm is exemplary. Projects should thus demonstrate a positive contribution to the surrounding environment, add to the social sustainability of both their immediate and wider settings, and represent design influenced by context, both environmentally and culturally.

Submissions can range in scale from a single site to a complete master plan of a neighborhood or city.

The site or master plan must be realized at the time of submission; proposals or visions are not eligible. In the case of a master plan, a multi-phase plan that is only partially completed will be accepted, but must be far enough progressed that its urbanistic intent is evident.

The "urban habitat" aspect of the project must have been completed and utilized in the two years prior to the current awards year, to have allowed time for the urban habitat to have proven its value (e.g., for the 2014 Awards, a project must have a completion date between January 1, 2012 and December 31, 2013).

Winner
Urban Habitat Award

The Interlace
Singapore

Completion Date: September 2013
Total Site Area: 60,000 sq m (645,835 sq ft)
Total Land Area: 81,000 sq m (871,877 sq ft)
Total Building Footprint: 21,000 sq m (226,042 sq ft)
Building Height: 89 m (292 ft)
Use: Residential
Owner/ Developer: CapitaLand Singapore Limited; Hotel Properties Limited
Architect: Office for Metropolitan Architecture, designer & partner-in-charge Ole Scheeren (now at Buro Ole Scheeren) (design); RSP Architects, Planners & Engineers Pte Ltd. (architect of record)
Landscape: ICN Design International Pte. Ltd. (design); Office for Metropolitan Architecture (concept)
Structural Engineer: TY Lin international
MEP Engineer: Squire Mech Private Limited
Main Contractor: Woh Hup Pte Ltd
Other Consultants: Acviron Acoustics Consultants Pte Ltd (acoustics); Langdon & Seah (quantity surveyor); Lighting Planners Associates (S) Pte Ltd (lighting)

"It is difficult to think of a more appropriate winner for the inaugural Urban Habitat Award than The Interlace. The project demonstrates brilliantly the real potential connected tall buildings have for delivering quality urban habitat at height."

Antony Wood, Juror, CTBUH

The Interlace is a 1,040-unit apartment complex consisting of 31 apartment blocks, each six stories tall and 70 meters long, stacked in hexagonal arrangements around eight large-scale, permeable courtyards. The stacking of the volumes creates a topographical phenomenon more reminiscent of a landscape than of a typical building. An extensive network of communal gardens and spaces is interwoven with amenities, providing multiple opportunities for social interaction, leisure and recreation – both on the roofs of, and in between, these stacked horizontal blocks.

Instead of following the default typology of housing in Singapore – clusters of isolated, vertical towers – the design generates an intricate network of living and social spaces integrated with the natural environment. The

> *"There was much interest in the project's success at forming structures for high-density living deployed horizontally, and the resulting benefit of outdoor urban environments."*
>
> Jeanne Gang, Jury Chair, Studio Gang Architects

blocks are arranged on four main "Superlevels" with three "peaks" of 24 stories. Other Superlevel stacks range from 6 to 18 stories to form a stepped geometry. The unusual geometry of the hexagonally stacked building blocks creates a dramatic spatial structure. Partly resting, partly floating, the blocks hover on top of each other to form open, permeable courtyards that interconnect with one another and the surrounding landscape and city. An expressive, interlaced space emerges that connects the multiple parts of the development into an open, inclusive community.

Multi-story openings between blocks allow light and air to weave into the architecture and through the landscape of the eight courtyards at the heart of the project. The primary pedestrian route through the project leads residents from the main entrance to the courtyards as primary points of orientation and identification – one lives in a courtyard, or a space, as opposed to a "building" or an "object." Pedestrian circulation is grouped and bundled according to the density of residents around each courtyard, by way of a central connector. A system of secondary footpaths brings residents on the most direct route from the connector to the front doors of their homes.

The landscape design capitalizes on the generous size of the eight-hectare site and further maximizes the green area and presence of nature. By stacking the apartment blocks, the design has generated additional horizontal surfaces, and thus the opportunity for extensive roof gardens and numerous landscaped public terraces,

Previous Spread

Left: Aerial view of complex

Right: Detail of stacked boxes with interstitial communal spaces

Current Spread

Above: View of towers looking up from base

Opposite Top: Site plan showing movement paths on ground plane

Opposite Bottom: Elevated walkways and public gathering spaces with relation to balconies

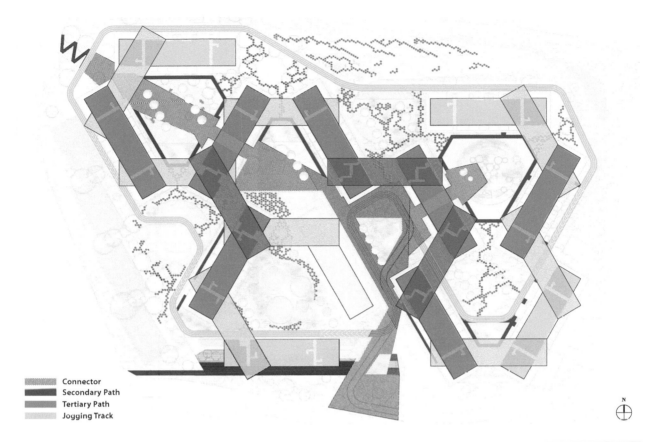

N

which, in aggregate, provide even more overall green area than the size of the unbuilt site.

A system of three core types, of 6, 18, and 24 stories respectively, is located at the overlap of the stacked apartment blocks. Cores typically serve three to four units per floor, which provides efficient circulation without long corridors. Core lobbies are naturally lit and ventilated, bringing daylight and fresh air into common areas. Circular "mega-columns" arranged around the vertical circulation in an optimized hexagonal configuration enable the three-way rotation of the blocks and provide a standard solution for all conditions.

The highly efficient system of compact cores, minimal circulation, and maximized floor area allowed the project to be realized on a budget for affordable private housing, within the competitive context of Singapore's market. A series of site-specific environmental studies – on wind, solar, and daylight conditions – were carried out and evaluated to determine intelligent strategies

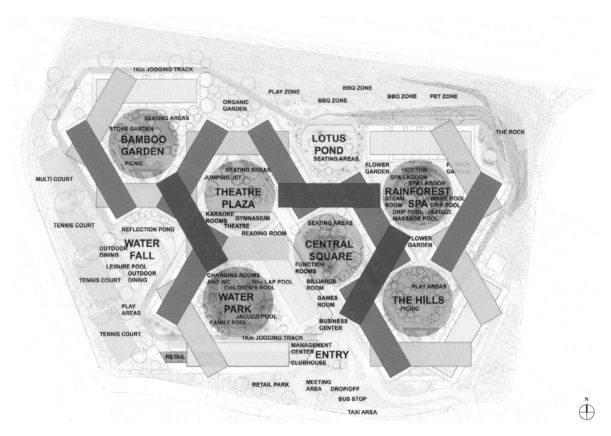

1Km JOGGING TRACK

PLAY ZONE

BBQ ZONE

BBQ ZONE

BBQ ZONE   PET ZONE

ORGANIC
GARDEN

SEATING AREAS

STONE GARDEN

**BAMBOO
GARDEN**

PICNIC

THE ROCK

**LOTUS
POND**

SEATING AREAS

MULTI COURT

SEATING AREAS

JUMPING JET

**THEATRE
PLAZA**

KARAOKE   GYMNASIUM
ROOMS   THEATRE
          READING ROOM

FLOWER   FLOWER
GARDEN   GARDEN
        SPA LAGOON
HOT TUB
        SPA LAGOON

**RAINFOREST
SPA**

STEAM   WHIRL POOL
ROOM
        DRIP POOL
DRIP POOL   JACUZZI
MASSAGE POOL

TENNIS COURT

REFLECTION POND

**WATER
FALL**

OUTDOOR
DINING

LEISURE POOL
        OUTDOOR
TENNIS COURT   DINING

SEATING AREAS

**CENTRAL
SQUARE**

FUNCTION
ROOMS

FLOWER
GARDEN

CHANGING ROOMS
AND WC   50m LAP POOL
        CHILDREN'S POOL

**WATER
PARK**

        JACUZZI POOL
FAMILY POOL

BILLIARDS
ROOM

GAMES
ROOM

PLAY AREAS

THE HILLS

PICNIC

PLAY
AREAS

BUSINESS
CENTER

TENNIS COURT

1Km JOGGING TRACK

RETAIL

MANAGEMENT
CENTER   **ENTRY**

CLUBHOUSE

RETAIL PARK

MEETING
AREA   DROP/OFF

BUS STOP

TAXI AREA

N

Top: Courtyard identity diagram displaying a multitude of public spaces

Left: Overall view of complex

Opposite: View of ground plane with amenities and tower core / "mega-column"

for the building envelope and landscape design. Early and comprehensive incorporation of low-impact passive energy strategies has won the project Singapore's Green Mark Gold[Plus] Award.

All apartments receive ample levels of daylight throughout the day, while the unique massing of the project provides a sufficient level of self-shading in the courtyards, which helps maintain comfortable outdoor spaces year-round for communal use.

Water bodies have been strategically placed within defined wind corridors. This allows evaporative cooling to happen along wind paths, reducing local air temperatures and improving the thermal comfort of outdoor recreation spaces in strategic micro-climate zones.

Extensive balconies and protruding terraces form a cascading vertical landscape across the façades and further connect the green roofs and shared communal terraces between the building volumes. Overall, the project appears not only surrounded by the tropical vegetation, but embedded within it.

All traffic and parking is accommodated in a single layer below the landscaped ground level. A large number of open-air voids allow light and air to reach the semi-sunken parking deck, creating areas of lush vegetation and trees below ground and connecting these spaces visually and through planting to the courtyards above.

Best Tall Building Asia & Australasia Finalist

In addition to being recognized as the inaugural Urban Habitat Award winner, the building itself has also been named a Finalist in the Best Tall Building Asia & Australasia category.

Jury Statement

Few projects so creatively realize the potential a tropical environment provides for inverting the "towers in the park" typology in favor of "towers as park" as does The Interlace. This integration of the best of horizontal and vertical living is far more than the sum of its parts. The scissoring, overlapping forms suggest innumerable possibilities for changing perspective, meeting new neighbors, or finding a longer way home, within one complex.

Taken apart from their stacked positions atop unseen axes, the relatively straightforward, balconied rectilinear forms reveal the immensity of past missed opportunities to orient International-Style regiment towards, or better yet, to render it part of the landscape. The plan view of the project reveals that it is not the arbitrary whim of a formalist at work; rather it is a logical outcome of a meticulous effort to maximize engagement with nature and with other humans, while still affording a sense of shelter, prospect, and mystery.

NEO Bankside
London, United Kingdom

Completion Date: October 2013
Total Open Area: 13,400 sq m (144,236 sq ft)
Total Site Area: 42,000 sq m (452,084 sq ft)
Total Building Footprint: 28,600 sq m (307,848 sq ft)
Building Height: 82 m (271 ft)
Use: Residential
Owner/Developer: GC Bankside LLP
Architect: Rogers Stirk Harbour + Partners
Landscape: Gillespies
Structural Engineer: Waterman Group
MEP Engineer: Hoare Lea
Project Manager: EC Harris
Main Contractor: Carillion PLC
Other Consultants: DP9 (planning); Gillespies (landscape); Hoare Lea (fire);
WT Partnership (cost)

"This project enhances public access to shared amenities, while stitching together the urban fabric with the bank of the Thames; a new experience is achieved in an ancient city."

Saskia Sassen, Juror, Columbia University

This residential scheme lies in the heart of the Bankside area of London, located close to the River Thames and directly opposite the west entrance to the Tate Modern and its new extension. Through careful arrangement of pedestrian pathways, landscaping, and building orientation, a generous public realm is created, which is animated by retail at ground level. Landscaped groves define two clear public routes through the site, which extend the existing landscape from the riverside gardens outside Tate Modern through to Southwark Street. They act as a catalyst for creating a lively and vibrant environment throughout the year.

The NEO Bankside development occupies a complex, irregular space with particular urban constraints, ranging from the large volume of the museum and its proposed extension, to the adjacent, listed two-story Almshouses. Moving the mass of the buildings away from Southwark Street, as well as from the Almshouses, helps mediate the difference in scale. The four individual buildings step up in height in response to the neighboring properties.

The grain of the development encourages permeability and public connectivity through the site, along a route that intuitively seems like the shortest, but also the most pleasant distance between two points. The resulting public realm extends the landscape from Tate Modern's own microcosm, creating a more consistent physical environment. A plot of land which housed the marketing suite and site office during construction will be gifted to Tate Modern, adding to the public space at the entrance.

The scheme incorporates a combination of renewable energy sources, which provide 10 percent of the energy requirements for the development. The design of NEO Bankside's energy/services strategy responds to Part L of the Building Regulations and the requirements of the Mayor of London's Energy Strategy. The development has achieved an EcoHomes "Good" rating.

NEO Bankside's four hexagonal towers have been arranged to provide residents with generous accommodation, stunning views and maximum daylight. The steel and glass pavilion-like buildings take their cues from the immediate context. The overall design hints at the former industrial heritage of the area during the 19th and 20th centuries, responding in a contemporary language which reinterprets the coloration and materials of the local architectural character. The oxide reds of the winter gardens echo those of Tate Modern and nearby Blackfriars Bridge, while the exterior's timber-clad panels and window louvers give the buildings a warm, residential feeling.

The design of the pavilions communicates permeability, particularly through the use of extensive full-height

"The hexagonal blocks of the towers are carefully worked into the triangular site, while the hard and soft landscape treatments unify the entire development."

Wai Ming Tsang, Juror, Ping An Development

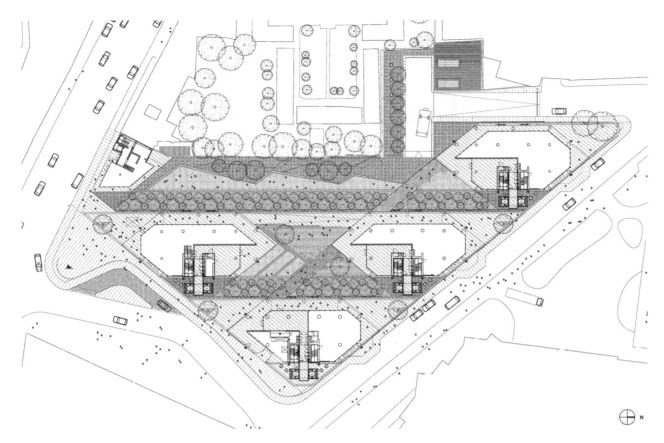

⊕ N

glazing and fully glazed winter gardens and transparent elevator cores. In addition, the placement of the buildings on the site helps to ensure that light is able to reach the central, landscaped gardens throughout the day at different times, despite the scale of the development.

NEO Bankside seeks to create a micro-ecological environment in an established urban setting by introducing an orchard of seasonal fruit trees, bat and sparrow boxes, a large and well-stocked herb garden and collection of active beehives, providing honey for residents. To enhance the surrounding social environment, a restaurant within the development encourages al fresco eating and drinking to create a vibrant, cosmopolitan atmosphere for all to enjoy.

Best Tall Building Europe Finalist

In addition to being awarded Finalist status in the Urban Habitat Award, the buildings themselves have also been named a Finalist in the Best Tall Building Europe category.

Previous Spread

Left: View of smallest tower as one weaves through the site

Right: Permeable ground plane between buildings

Current Spread

Opposite Top: Public green space between towers

Opposite Bottom: Sculpture along crossing pathways at tower base

Above: Site plan showing four hexagonal pavilions and interconnecting landscaped public pathways

Jury Statement

Relief must have rippled through Southwark, a dense riverside district with historic and contemporary connections to the waterfront, when it was realized that not only did this tall building project not block views, it actually enhanced access to the waterside and provided a new logical right of way for pedestrians that is as pleasant as it is direct. Both the hard and soft landscape treatments unify the entire development.

Gramercy Residences, SkyPark
Makati, Philippines

Appropriate conditions for high-rise living in the tropical climate, coupled with a shortage of open social space in Makati City, allowed the designers to introduce a recreational social space that caters to the building's residents, as well as visitors, in the form of a two-level Sky Park at the 36th and 37th stories. Recreational facilities include a gym, swimming pools, bar, library, crèche, movie theater, and spa/wellness areas, within a garden-like setting that seeks to improve comfort levels and moderate the immediate micro-climate.

Landscaped sky courts and sky gardens, particularly when deployed at a critical activity center midway up the tower, can encourage people to spend more time outdoors, undertaking social and recreational activities and thus heighten the likelihood of social interaction through chance meeting. The horizontal greenery of the original Gramercy Park in New York has been extrapolated and rotated 90 degrees to effectively create a green wall "mural" within a contemporary vertical community on the other side of the planet.

Completion Date: December 2013
Floor area for skypark: 3,500 sq m (37,674 sq ft)
Building Height: 244 m (800 ft)
Use: Residential
Owner/Developer: Century Properties Group, Inc.
Architect: Jerde Partnership Inc.; Pomeroy Studio Pte Ltd
Landscape: Pomeroy Studio Pte Ltd
Structural Engineer: Arup
Main Contractor: Consis Engineering (for SkyPark Greenery)

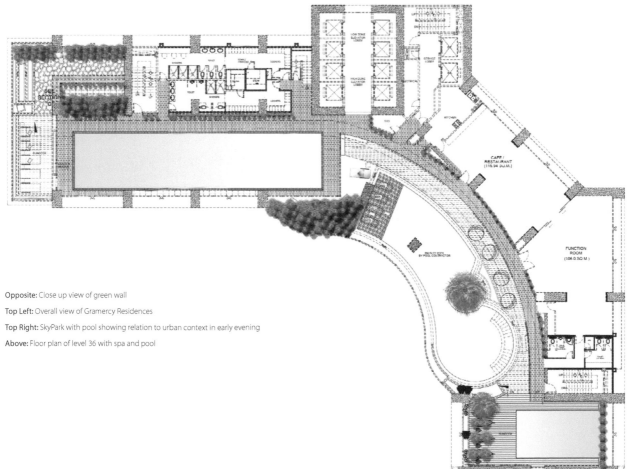

Opposite: Close up view of green wall

Top Left: Overall view of Gramercy Residences

Top Right: SkyPark with pool showing relation to urban context in early evening

Above: Floor plan of level 36 with spa and pool

CTBUH 10 Year Award

Award Criteria

The CTBUH "10 Year Award" recognizes proven value and performance (across one or more of a wide range of criteria) over a period of time. This award gives an opportunity to reflect back on buildings that have been completed and operational for at least a decade, and acknowledge those projects which have performed successfully long after the ribbon-cutting ceremonies have passed.

Evidence must be provided of performance in any category, including but not limited to: contribution to urban realm, contribution to culture/iconography, social issues, internal environment, occupant satisfaction, technical/engineering performance, environmental performance, energy performance, etc.

For a building to be considered for the CTBUH 10 Year Award, it must have been completed 10–12 years prior to the current award year (e.g., for the 2014 award, a project must have a completion date between January 1, 2002 and December 31, 2004).

Winner
10 Year Award

Post Tower
Bonn, Germany

Completion Date: 2002
Height: 163 m (533 ft)
Stories: 42
Area: 107,000 sq m (1,151,738 sq ft)
Use: Office
Owner: Deutsche Post Bauen
Developer: Deutsche Post AG
Architect: JAHN (design); Heinle Wischer + Partners (architect of record)
Structural Engineer: Werner Sobek Group
Energy Concept: Transsolar
MEP Engineer: Brandi Consult GmbH
Main Contractor: HOCHTIEF Construction AG Niederlassung Hamburg
Other Consultants: L-Plan (lighting); Peter Walker (landscape)

"The design single-handedly changed the way we look at façade and office design to date. It has become an exemplar sustainable icon in contemporary architecture."

David Gianotten, Juror, OMA

Deutsche Post DHL's Bonn headquarters is an expression of the company's confidence in progress and technology to improve the work environment. The architect's aim was to create a forward-looking high-rise for the 21st century: formally, technically, and ecologically advanced, and offering a high-quality work environment.

The Post Tower demonstrates the potential for technically integrated buildings to deliver high performance. From the onset of planning, the client expressed a strong desire to give all office staff direct access to outside air and natural light. Combining these factors led the design team to develop the twin-shell façade and split-building typology that embraces natural ventilation and green energy sources.

The building uses natural ventilation and decentralized cooling/heating systems to reduce the mechanical load, and consequently energy consumption, while improving floor-plate efficiency and eliminating

Jury Statement

The Post Tower undoubtedly changed the way we look at sustainable building for the high-rise typology. The building paved the way for many successors, crossing the mental barriers many designers, consciously or not, had thrown into their own way.

A building shrouded in glass can indeed work to reduce energy load. It is not necessary to affix visibly active technology to a building's surface in order to communicate its functionality. While precisely designed, the double-façade solution is not technically complex, therefore increasing its practicality and longevity, both for this building and for those that would follow.

The fact that user behavior subverted some of the performance goals for the building simply demonstrates that Post Tower was ahead of its time, and we as a building culture, designers and occupiers alike, still have much to learn.

Previous Spread

Left: Overall view of the tower in context

Right: Detail view of the twin shell façade with natural ventilation features

Current Spread

Left: View of atrium which runs the entire height of the building, divided into four parts

Opposite Left: Detailed section through the south façade showing operable flaps to let air into the cavity

Opposite Right: Typical floor plan

mechanical shafts and ceiling ducts. This allows the tower to consume only 75 kWh/m^2 (measured through 2003), which represents a 79 percent energy reduction when compared to a typical air-conditioned building. Only 3 kWh/m^2 are used for cooling, and the energy used for heating is drastically reduced compared to a benchmark building. These savings in energy usage and central air-conditioning system equipment and space offset the additional investment in the façade.

The Tower's form consists of two offset, elliptical segments separated from each other by a 7.2-meter-wide atrium that faces west towards the City of Bonn, and east towards the Rhine River. The atrium runs the full height of the Tower, incorporating the glazed passenger elevators, and it is divided into four parts by sky gardens; three are nine stories high and the top one is eleven. In each elliptical segment, cellular offices wrap around the perimeter, with conference rooms and core functions located towards the atrium. The two main façades of the high-rise face north and south, while the sky garden façades have east and west orientations. By orienting the building in the primary wind direction, the wind profile was optimized, both for structural efficiency and natural ventilation.

Outside air enters the building through the twin-shell façade, flows through the offices and into the corridors, which act as horizontal exhaust air collectors, ultimately venting the exhaust air into the atrium. Using the stack effect, the exhaust air in the atrium is then vented through operable windows

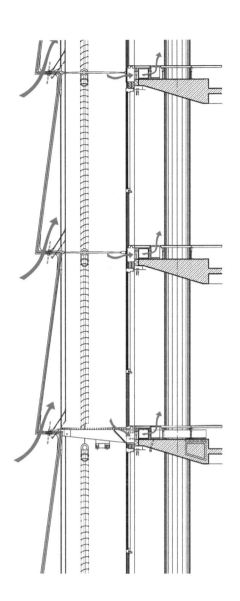

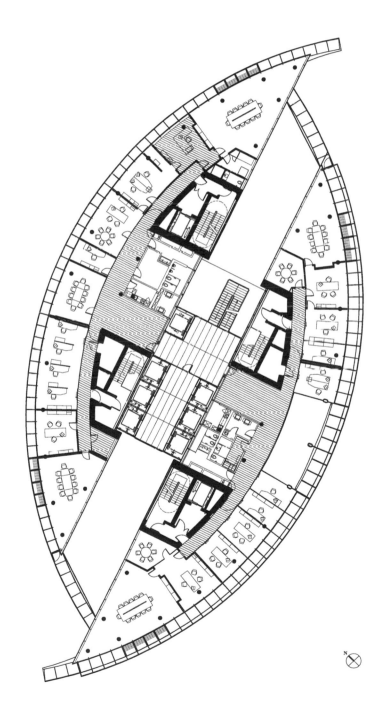

located high in the façade of each nine-story atrium, with low-level vents added to assist the natural airflow.

The façade consists of the outer single-glazed façade, operable sunshades, and an inner façade of floor-to-ceiling insulated glazing. The horizontal continuity of the façade and its aerodynamic shape allow it to dissipate the pressure differences across the faces of the building, enabling natural ventilation to take place without a draft.

The outer façade is hung in nine-story increments from extruded stainless steel mullions, and braced horizontally at every floor with wind needles. The south façade features sloped glass panes, allowing air-intake and exhaust at the bottom of each panel. The north façade is a smooth plane with alternating ventilation flaps.

The inner façade is a floor-to-ceiling aluminum curtain wall with insulated glazing. Motorized, operable

windows are located in every other façade module. During spring and fall, these windows serve to naturally ventilate the offices, substantially reducing the building's reliance on mechanical conditioning. After working hours, the windows are centrally opened to provide night flushing of the tower with cool air.

Each employee can manually alter the internal environment according to individual preferences using a touch-screen, allowing individual control of the blinds, lighting levels, operable windows, and internal temperatures.

The design of the Post Tower opened up new possibilities for the work environment, promoting interaction and communication, and delivering flexible spaces that can accommodate new layouts and technologies.

"The tower was an early touchstone for sustainable tower design and has provided a very useful model through its attention to performance. The jury was impressed with the continued monitoring and proof of concept."

Jeanne Gang, Jury Chair, Studio Gang Architects

Taipei 101
Taipei, Taiwan, China

Completion Date: 2004
Height: 508 m (1667 ft)
Stories: 101
Area: 198,347 sq m (2,134,989 sq ft)
Use: Office
Owner/Developer: Taipei Financial Center Corporation
Architect: C.Y. Lee & Partners Architects/Planners
Structural Engineer: Evergreen Engineering (design); Thornton Tomasetti (peer review)
MEP Engineer: Continental Engineering Consultants, Inc. (design); Lehr Associates (peer review)
Main Contractor: Kumagai Gumi; RSEA Engineering; Samsung C&T Corporation; Ta-You-Wei Construction; Taiwan Kumagai
Other Consultants: ALT Cladding (façade); Genius Loci (landscape); Lerch Bates (vertical transportation); RWDI (wind); Shen Milsom Wilke, Inc. (acoustics); Siemens Building Technology (energy concept); SL+A International Asia (LEED)

Torre Agbar
Barcelona, Spain

Completion Date: 2004
Height: 144 m (474 ft)
Stories: 35
Area: 47,500 sq m (511,286 sq ft)
Use: Office
Architect: Ateliers Jean Nouvel (design); b720 arquitectos (architect of record)
Structural Engineer: Robert Brufau y Asociados (design); Obiol, Moya i Associates (engineer of record)
MEP Engineer: Axima Sistemas e Instalaciones S.A.; Gepro S.A.
Main Contractor: Dragados

Uptown Munchen
Munich, Germany

Completion Date: 2004
Height: 146 m (479 ft)
Stories: 38
Area: 50,200 sq m (540,348 sq ft)
Use: Office
Developer: Gerald D Hines Interests
Architect: ingenhoven architects
Structural Engineer: bwp Burggraf + Reiminger Beratende Ingenieure GmbH
MEP Engineer: Ingenieur Consult
Other Consultants: BPK Brandschutzplanung Klingsch (fire); DS-Plan GmbH (acoustics, façade); Ingenieurgesellschaft Heimann (traffic); Kardorff Ingenieure (lighting)

Highlight Towers
Munich, Germany

Completion Date: 2004
Height: 126 m (413 ft)
Stories: 33
Area: 73,189 sq m (787,800 sq ft)
Use: Office
Architect: JAHN
Structural Engineer: Werner Sobek Group
MEP Engineer: ENCO Energie-Consulting GmbH
Main Contractor: STRABAG AG

Time Warner Center

New York City, United States of America

Completion Date: 2004
Height: 228 m (749 ft)
Stories: 55
Use: Residential/Hotel/Office
Developer: Apollo Real Estate Advisors, L.P.; Columbus Centre LLC; The Related Companies
Architect: Skidmore, Owings & Merrill
Structural Engineer: WSP Cantor Seinuk
MEP Engineer: Cosentini
Main Contractor: Lend Lease
Other Consultants: Langan Engineering & Environmental Services (geotechnical)

Bloomberg Tower

New York City, United States of America

Completion Date: 2004
Height: 246 m (806 ft)
Stories: 54
Area: 84,000 sq m (904,168 sq ft)
Use: Residential/Office
Architect: Pelli Clarke Pelli Architects (design); Adamson Associates (architect of record); SLCE Architects (architect of record)
Structural Engineer: Thornton Tomasetti
MEP Engineer: WSP Flack + Kurtz
Main Contractor: Lend Lease
Other Consultants: RWDI (wind)

Tower Palace Three
Seoul, South Korea

Completion Date: 2004
Height: 264 m (865 ft)
Stories: 73
Use: Residential
Architect: Skidmore, Owings & Merrill (design); Samoo Architects & Engineers (design)
Structural Engineer: Skidmore, Owings & Merrill (design); Samoo Architects & Engineers (design)
MEP Engineer: Samsung C&T Corporation

Award Criteria

This award recognizes a specific area of *recent* innovation in a tall building project that has been incorporated into the design, or implemented during construction, operation, or refurbishment. Unlike the CTBUH Best Tall Building awards, which consider each project holistically, this award is focused on one special area of innovation in the project – thus not the building overall. The areas of innovation can embrace any discipline, including but not limited to: technical breakthroughs, construction methods, design approaches, urban planning, building systems, façades, interior environment, etc.

The important criteria for judging is that the submission outlines succinctly the area of innovation, in comparison to standard benchmarks. The Innovation award can include recognition of a breakthrough that may not yet have been implemented in a specific building, but has been thoroughly tested.

The project must clearly demonstrate a specific area of innovation within the design and/or construction that is new and pushes the design of tall buildings to a higher level. The area of innovation should demonstrate an element of adaptability that would allow it to influence future tall building design, construction, or operation in a positive way.

Winner
Innovation Award

BioSkin

Innovation Design Team:
Nikken Sekkei

Left: Detailed view of BioSkin façade

Above: Overall view of NBF Osaki Building, Tokyo, where BioSkin has been implemented

"This innovation is a bold concept, which has been suitably analyzed, elegantly integrated into the architectural form, and beautifully detailed in its execution."

Paul Sloman, Technical Juror, Arup

As it has in many other cities, the annual average temperature of metropolitan Tokyo has risen 3°C in the past 100 years due to the increase in built-up hard surfaces, otherwise known as the "urban heat island effect." As a response to this, a conceptually simple system has been devised, based on the traditional Japanese practice of *uchimizu;* the sprinkling of water to lower ambient temperatures, clean the streets and keep dust at bay.

BioSkin is a system of ceramic pipes, affixed to the side of a building, which absorbs heat through causing evaporation of the rainwater inside the pipes, mitigating the urban heat island effect by cooling the building as well as its immediate surroundings. Through this process, the surface temperature of the building enclosure can be reduced by as much as 12°C and its micro-climate by about 2°C. The potential implications of this are substantial: if a large number of buildings in a city used such a system, ambient air temperature could be reduced to the point that cooling loads for many

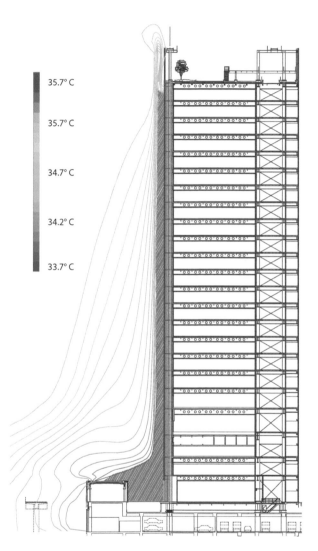

35.7° C

35.7° C

34.7° C

34.2° C

33.7° C

"Bioskin allows rainwater recycling to be further utilized for a reduction of building energy through its carefully designed façade system."

Douglas Mass, Technical Juror, Cosentini Associates

pipes, which in the live test case were incorporated as balcony railings on a Tokyo office building, reminiscent of the horizontal screens seen throughout Japan and known as sudare. Rainwater penetrates outward through the porous ceramic, evaporating from the pipe's surface, cooling the surrounding air. Excess water is then drained down to the soil of the premises to the extent possible, normalizing the water cycle and reducing the load on sewage infrastructure.

To assure the system's operational effectiveness, a mock-up experiment was conducted in the summer of 2008. To ensure an environment as close as possible to an actual installation condition, the system was arranged to face the northeast side of the building, shielding the façade from solar radiation where it would be most concentrated. The surface temperature and atmospheric data were measured, and the loss of water in the ceramic pipe was regularly measured to calculate the volume and rate of evaporation. The simulation demonstrated that the surface temperature of the water-retentive ceramic pipe was lowered by up to 10°C; airflow analysis confirmed that the temperature of the surrounding air decreased by up to 2°C.

Perhaps the highest praise for BioSkin is drawn from the fact that it is a locally devised solution that fuses an age-old custom with a prevalent vernacular design strategy, to achieve an outcome that could have wide-ranging, even global implications.

buildings, even those without the system installed, could be reduced.

The simplicity of the system is elegant. The BioSkin tubes are made of extruded aluminum cores, with a highly water-retentive terra-cotta shell attached to the aluminum core using an elastic adhesive. When rainwater collects on the rooftop, it is then drained to a subsurface storage tank, where it is filtered and sterilized. This water is then pumped up and circulated through the

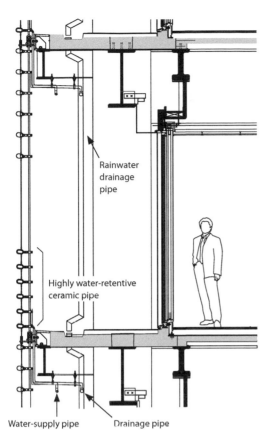

Rainwater
drainage
pipe

Highly water-retentive
ceramic pipe

Water-supply pipe   Drainage pipe

Opposite: Section of building detailing effects BioSkin has on the micro-climate around the façade

Left: Detail section of BioSkin façade

Right: Louver section

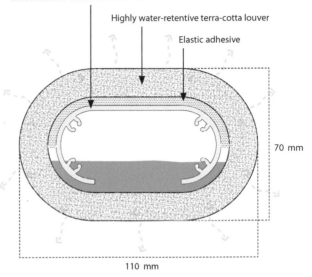

Core material: Aluminum extruded material

Highly water-retentive terra-cotta louver

Elastic adhesive

70 mm

110 mm

There is great potential in such a system, not only for meaningfully driving down energy consumption in their host buildings and in those that surround them, but also for supporting the water ecology of the local area. The filtered water that does not return to the atmosphere with the attendant cooling effect instead returns to the ground, perhaps cleaner than when it fell from the sky.

This water irrigates the soil, while avoiding washing oil slicks and other hazardous material associated with paved areas into the sewer system and ultimately, the sea. Of course, to be truly effective, such systems need to be reproduced at larger scale. Through the publicity developed through awards programs such as this one, it is hoped that such systems can be economically mass produced for buildings in all kinds of climates, while incorporating themselves into appropriate architectural variations along the way.

Jury Statement

By going back to roots and merging the traditional practice of *uchimizu* with the time-tested vernacular typology of horizontal screens for solar shading, the BioSkin team has delivered in this demonstration an outstanding project of individual architectural integrity, while bringing to light a highly useful, repeatable concept that could have a real impact on the climates of our future cities. Its exploitation of relatively simple technologies that have been around years, and in some cases, centuries, makes this a particularly important rejoinder to those who would argue that our urban and environmental problems can only be tackled with punitively expensive and time-consuming undertakings that will undermine economic progress. As if urban-heat-island-mitigation weren't enough, BioSkin also presents a plausible remedy for overloaded sewer systems and marine pollution.

All of the natural processes that make BioSkin work – gravity, rainwater, envirotranspiration, and water pressure – have been available to humans since the dawn of our existence on this planet. After we finish asking, "Why didn't we think of this before?" the only question that remains is, "How soon can we implement this widely?"

Active Alignment for Tall Buildings with Unusual Geometry

"This method provides an effective and simple way to compensate for uncertainty and variability of on-site conditions for buildings with unusal shapes."

Nengjun Luo, Technical Juror,
CITIC HEYE Investment CO., LTD.

In any tall building with asymmetric loading, there is increased potential for the building to lean to one side. This was the case on the iconic Leadenhall Building, a 224-meter-tall office building located in The City of London. The awardees developed an alignment method for bolted connections in steel, which has been proven in practice in the construction of the Leadenhall Building. Between the flank diagonal columns there is a set or "pack" of temporary plates that act as adjustable

Innovation Design Team
Laing O'Rourke

Skidmore, Owings & Merrill; Arup (structures)
Watson Steel Structures Ltd (Steel Manufacturer)
Cambridge University (Monitoring Technique Research)

Left: The Leadenhall Building, London, during construction, where the active alignment strategy was implemented

Opposite Top: Example of a diagonal member, using the active alignment strategy

Opposite Bottom: A temporary packer plate is removed during construction

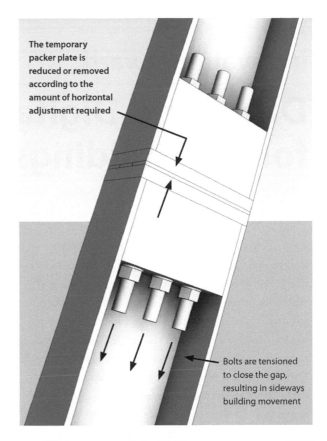

The temporary packer plate is reduced or removed according to the amount of horizontal adjustment required

Bolts are tensioned to close the gap, resulting in sideways building movement

shims, which can be removed or added as needed, by opening the joints vertically with hydraulic jacks, and horizontally by tightening or loosening the bolts.

The sequence of stressing is predicted by the use of staged finite element analysis, and adjusted to ensure the continuous verticality of the building. As a result, erection begins straight and ends straight; thus, manufacturing details and site lay-down are greatly simplified. As the pack adjustment process is flexible, the number of shims and jacking distances can be easily modified so that uncertainties and variability in the finite element analysis, site progress, and site conditions are readily compensated for.

Where the permanent works design permits, elements of the design can be used as stressing elements to act as "turfers" to ensure vertical alignment. The cost of such an approach is thus minimized, as there is little design complexity in details or manufacturing, and a high level of analytical accuracy is not required, relative to the amount of vertical alignment accuracy achieved.

The active alignment methodology could be developed further to remain active during building operation, so that longer-term movements, such as geotechnical settlements, could be corrected. The solution could be enhanced further to respond to building movements associated with wind and earthquakes in real time, enabling the construction of taller and more slender buildings, which would in turn maximize the efficient usage of urban space.

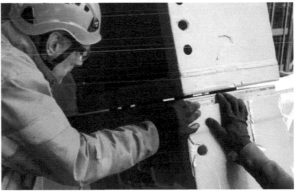

Jury Statement

As buildings get taller and their shapes become more complex, these engineers have devised a structural system that allows simple adjustments which ensure structural tolerances are met both in the short and long term. The engineering solution is a direct response to real construction issues and building movement over time.

DfMA and Digital Engineering for Tall Buildings

As building geometries are becoming more complex and the industry builds higher, swifter, and safer, higher-quality construction techniques are required. This is at odds with the prevailing psychology of the construction industry, where projects are awarded to the lowest-price bidder, which often involves the greatest underestimation of the task at hand, and the most risk-accretive approach. With this process flowing down the supply chain, design decisions are delayed until the last possible moment, hampering the ability to pre-plan or conduct off-site manufacturing. However, as procurement methods become more collaborative and clients engage earlier in the process, and as government targets for CO_2 reduction are tightened, research and investment in new construction processes has moved forward.

The DfMA (Design for Manufacturing and Assembly) and Digital Engineering approach to tall building clearly addresses the ever-increasing challenges normally associated with tall buildings, of congested urban environments, increasing architectural complexity, and greater demands on speed, quality, safety, and sustainability. The approach emphasizes early collaboration between disciplines through shared digital models, and project-based, multidisciplinary special teams for complex sections of the project.

The method has been proven in practice in the construction of the Leadenhall Building, a 224-meter-tall office building located in The City of London. In this project, more than 80 percent of the project's total value was built off-site using the approach.

For example, a separate project team was established just to deliver the building's "attic," containing most of its mechanical plant, with months of detailed pre-planning and coordination that included the commissioning process. Some of the outputs that make their way into the field include "method statements," which provide specific, diagrammatic instructions for assembling pre-fabricated parts.

In the demonstration project, prefabricated elements included: the steelwork frame, precast concrete planks and core elements, unitized façade panels, panelized bathrooms, modular services risers and plant rooms, pre-assembled escalators and elevator cars, and core "tables," consisting of the integrated delivery of steel tables for the building core with pre-loaded concrete floors and pre-installed ceiling services. These were accompanied and guided by integrated 3D models of the structure, services, cranes, and façade.

All of these contributed to the success of the project, and have likely application for many other tall buildings, especially in other congested urban conditions.

Innovation Design Team
Laing O'Rourke

Arup (structures, MEP engineer)
Rogers Stirk Harbour + Partners (design architect)

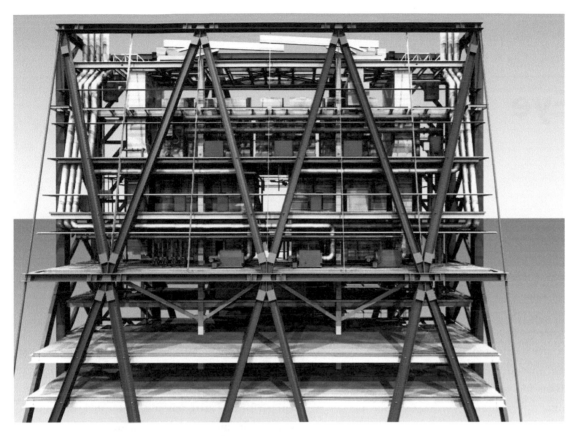

Top: Digital model of the Leadenhall Building's "attic"; carefully designed to accommodate most of the building's mechanical plant

Left: Digital model of the complex services installation

Right: Prefabricated modules arriving on-site to be carefully fit into place according to the digital models

Nominee
Innovation Award

LiftEye

LiftEye is the world's first "virtual elevator window system." LiftEye was designed to enhance the elevator passenger's experience by converting a conventional elevator with opaque walls into a "panoramic" elevator, by transmitting real-time images from the building's exterior that correspond to the floor being passed by the cabin. The spectator's point of view corresponds to actual current elevator car height. LiftEye can function as a basis for various entertainment services such as streaming information to further enhance the passenger experience.

The system fits up to three HD LCD widescreen monitors in a column on the car wall. It functions due to a small number of specifically designed cameras fixed outside of the building and through software, which transforms the building environment in real time, adding depth and details, such as raindrops rolling on the screen when it is raining outside, thus replicating the panoramic elevator experience at a fraction of the cost. The system is designed to be applied to both new and pre-existing elevator installations and is adaptable to any elevator model, and it does not interfere with elevator safety systems. The software can accept third-party connections for additional content displays, such as "augmented reality," advertising, or newsfeeds. Video recording is excluded, resolving any privacy violation concerns.

The high-definition displays can also be connected across buildings allowing them to show panoramic views broadcast from any other building in the world that has a LiftEye system in place. For example, a hotel chain with a high-rise in Shanghai could display real-time panoramic views of properties in New York and London, simultaneously in one car. Further commercial possibilities are being explored.

Several trial installations have already taken place. In the Old Water Tower at the Water Museum in St Petersburg, Russia, in early 2013, the development team performed staff training, and tested the overall structure, the means of data transmission from exterior cameras to a moving elevator car, and the performance of software and hardware on specific models of video camera. There was also a temporary installation at the Augsburg Messe Exhibition Center in Germany, during the Interlift biannual trade fair in October 2013. Several commercial projects involving LiftEye in the United States, Europe, and the Middle East are currently in the pre-construction phase with official announcement being made early 2015.

Innovation Design Team:
LiftEye Ltd

Stein Ltd (Elevator Design and Maintenance)
Mathematics and Mechanics Faculty of St Petersburg State University (Software Design)
LM Liftmaterial GmbH (Installation)

Left: Rendering of elevator featuring LiftEye technology

Opposite Top Right: LiftEye camera

Opposite Middle Right: Demonstration of LiftEye technology

Opposite Bottom Right: Potential future LiftEye installation

Steel Fiber Reinforced Concrete

While steel fibers are commonly used in tunnel linings, industrial floors, and other applications where high toughness is required, their use in building structures has previously been limited.

Recently, however, steel fiber reinforced concrete (SFRC) was used as a substitute for seismic reinforcing bar in shear-wall link beams, eliminating all link beam diagonal bars and significantly reducing the quantity of remaining rebar.

The new application involves mixing high-strength steel fibers into the link beam concrete. The SFRC eliminates the unwieldy diagonal bars and much of the remaining steel that is otherwise required for conventional link beam designs in regions of high seismicity, providing improved strength and ductility. Further, SFRC mitigates the serious constructability challenges of placing rebar and concrete in what are typically highly congested zones.

The Martin in Seattle, Washington, a 24-story apartment tower over subterranean parking, served as the inaugural project for shear-wall seismic design using SFRC. The structural system is cast-in-place concrete with post-tensioned slabs. In response to demanding seismic requirements, the structure incorporated a heavily reinforced shear-wall core. As a result of openings through the core walls, the core included link beams. The link beams play a critical structural role in connecting the wall piers, particularly during an earthquake.

The most difficult and congested area of a reinforced concrete structure is often in these shear-wall link beams. Traditionally, diagonal bars are used to reinforce link beams in high seismic regions, combined with large quantities of tightly spaced stirrups and ties. This traditional approach is costly and time consuming to build, however, and creates major congestion, since the diagonal bars conflict with rebar in the adjacent wall boundary elements.

With SFRC, the strength and ductility of link beams is maintained, even though a significant quantity of reinforcing steel is eliminated. SFRC can result in up to a 40 percent rebar reduction and a 30 percent net cost savings as compared to traditional link beam construction.

This innovation saves significant labor and material, and provides the structural engineering profession with a new and valuable tool for improving the design of reinforced concrete buildings in high seismic regions.

Innovation Design Team:
Cary Kopczynski & Company (structural engineer)

Vulcan Real Estate (Owner/Developer)
Callison (Architects)
Exxel Pacific (General Contractor)
Conco (Concrete Contractor)
Cadman (Ready Mix Supplier)
Bekaert (Steel Fiber Supplier)
University of Michigan Civil and Environmental Engineering (Researchers)

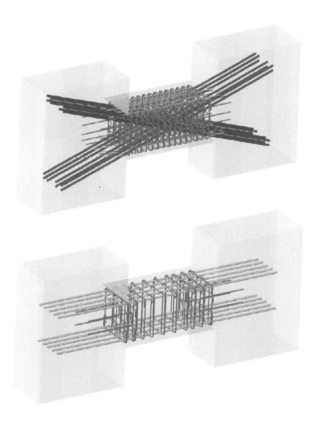

Left: Conventional reinforced concrete coupling beam (top) and the SFRC coupling beam (bottom) showing how SFRC can significantly reduce material quantities and simplify construction.

Top Right: Sample of steel fiber reinforced concrete (SFRC)

Bottom Right: The Martin, Seattle, where the SFRC system was implemented in shear-wall link beams

CTBUH Performance Award

Award Criteria

The CTBUH Best Tall Building awards, like most awards programs, recognize new buildings – based partly on the stated intentions of these buildings. It is increasingly being recognized that the industry needs to focus on actual "performance" rather than "best intentions" and thus, the CTBUH "Performance" award recognizes proven environmental value and performance over a period of time.

The Performance Award recognizes the measured environmental performance of a building or development and the award goes to the building that has the least environment impact on the urban realm using measured data. The purpose of the Award is to recognize buildings that excel in their particular condition.

The collection and presentation of key data is required and the building must have more than a year of measured data.

Environmental performance can be assessed in any category, including but not limited to: internal environmental performance, energy use, energy improvements (e.g., reductions for refurbishment), use of natural resources, energy creation, carbon footprinting, etc. The performance should be linked to occupant satisfaction and contribution to urban realm although this will be subjective and shall only be a deciding factor between buildings of similar performance.

For a building to be considered for the CTBUH Performance Award, it must be a minimum of three years since completion and the building should have 70 percent occupancy or above.

Winner
Performance Award

International Commerce Centre

Hong Kong, China

Completion Date: 2010
Height: 484 m (1,588 ft)
Stories: 108
Area: 273,148 sq m (2,940,167 sq ft)
Use: Hotel/Office
Owner/Developer: Sun Hung Kai Properties
Architect: Kohn Pedersen Fox Associates (design); Wong & Ouyang (architect of record)
Structural Engineer: Arup
MEP Engineer: J. Roger Preston Group; Parsons Brinckerhoff
Project Manager: Harbour Vantage
Main Contractor: Sanfield Building Contractors Limited
Management Company: Kai Shing Management Services Limited
Other Consultants: Belt Collins & Associates (landscape); Lerch Bates (vertical transportation); Permasteelisa Group (façade); RWDI (wind); W T Partnership (cost)

"The ICC's management team gives a shining example, their year-on-year progress clearly recognizes the importance of interaction with their tenants and the community."

Peter Williams, Technical Juror, AECOM

The International Commerce Centre is the tallest building in Hong Kong, housing some of the most prominent financial institutions in the world, at 98 percent occupancy. The building is routinely recognized as a paragon of good management, from a commercial, environmental, and community standpoint.

The level of energy efficiency achieved by the ICC is unusual for a tall building, and significant investments have been made in improving energy performance over the years, especially since adopting the ISO 50001 Energy Management Systems certification in 2011. This commitment was followed by more than 50 advanced energy-saving measures. The Energy Utilization Index (EUI) of ICC's energy performance in 2013 was 157.3 kWh/m^2, placing it among the top 10 percent of energy-efficient commercial buildings. A computerized building management system manages and controls the energy use in the building. The total energy consumption of the project was reduced from 56.3 million kWh in 2012 to 49.9

"The ICC owners and management team have made an outstanding commitment to describing and documenting the building's performance, looking at sustainability in a comprehensive and holistic way. Their documentation is exemplary and sets a new bar for clarity about energy use."

David Scott, Technical Jury Chair, Laing O'Rourke

million kWh in 2013, a reduction of 6.4 million kWh, or 11 percent.

The air-conditioning system is a high-voltage water-cooled chiller system with a centrifugal separator enhancing the chiller's coefficient of performance (COP), resulting in an 8 percent reduction in energy consumption. The system's original corrugated aluminum separator box filter was replaced with a more advanced mini-pleat filter, reducing system pressure by 25 percent and consequently reducing energy consumption as well. Management carried out a life cycle testing program with Hong Kong Polytechnic University, resulting in energy optimization and significant savings of 7 million kWh from 2011 through 2013. ICC has saved an estimated HK$7 million (US$900,000) annually through this program.

Other energy-saving features include a low-emission curtain wall, natural lighting of the atrium, the wide adoption of energy-efficient lighting fixtures such as LEDs and T5 fittings, and double-decker elevators with destination control and power regeneration functions. Simple actions such as deactivating elevators during low-use periods are equally important. Through these investments, the building reduced CO_2 emissions by 4.2 million kilograms in 2013.

The building's waste-management program has increased its collection of recyclables from 94,000 kilograms in 2012 to 101,714 kilograms in 2013, an 8.21 percent increase. Participation in the recycling program increased

Previous Spread

Left: Overall view of tower

Right: ICC Help Desk in operation – one of the measures used to maintain an involved relationship with the building's tenants

Current Spread

Opposite: Graph comparing the building's annual energy consumption

from 60 percent of occupants in 2010 to 90 percent in 2013. Recycled items collected have expanded to include waste paper, plastic and aluminum cans, glass bottles, food waste, chemical waste, and coffee grounds, as well as creative reuse of festival decoration items.

Of course, a building's initial successes in performance-based design can be undone if occupants are not well trained and motivated to continue imbuing their daily activities with these principles. The key to achieving high performance lies in ICC's intensive, involved relationship with its tenants. Each tenant is assigned its own account manager, who is in frequent contact with the tenant and learns about their requirements in detail. A 24/7 helpdesk is available for tenants, regardless of where they are in the world.

To ensure ongoing "greenkeeping" by occupants and the wider community, ICC's management extended its food-waste collection program to the surrounding community, and conducted more than 300 sharing

Annual Energy Consumption in 2012, 2013 and 2014 (through March)

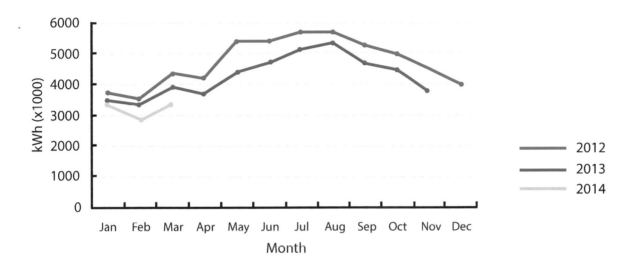

events to promote green principles. The "ICC Celsius 26" plan was launched in 2012, maintaining the building's common areas at 26 degrees Celsius, warmer than typical engineering standards, but still comfortable. Apart from the base building's energy audit, ICC provides free energy audits for its tenants upon request.

By using the helpdesk and computerized building systems as a data-collection and dissemination tool, combined with a high level of direct training, interaction and support of tenants, a "virtuous circle" is formed, in which tenants are motivated to adhere to energy guidelines and to contribute data, which is then reported back to tenants and management.

Jury Statement

The International Commerce Centre is one of the most energy-efficient tall buildings currently in existence, and is also one of the most well run. This superlative was achieved partially by way of wise technology investments in mechanical equipment, façades, and lighting, and partly by way of direct engagement with tenants and encouragement of building stewardship. The history of the ICC is a consistent battle to improve performance, by tactically replacing outdated and inefficient equipment, actively testing new approaches with academic researchers, and by building programs and policies that ensure everyone in the building views it as a shared asset and a shared responsibility.

At ICC, the key to enhancing the performance of man, machine, and building in an integrated, productive fashion was data transparency. By collecting, logging, and sharing progress on dozens of energy-saving measures, including those undertaken by tenants, a collective sense of commitment and competitiveness was engendered. The operational management structure of this building is truly amazing, demonstrated by a commitment to improve all aspects of its performance incrementally, year on year.

Finalist
Performance Award

Jin Mao Tower
Shanghai, China

"This project adopted many innovative measures which show a strong commitment to energy and resource savings. It sets a new standard for high-rise buildings in China."

Guo-Qiang Li, Technical Juror, Tongji University

Jin Mao Tower, a mixed-use complex containing offices, convention space, and a hotel, was built in 1999, and in 2013 became the tallest and the longest-operated building in mainland China to receive a LEED-EB: OM (Existing Buildings: Operations + Management) Gold certification. Its high performance has been achieved with the assistance of a computerized energy management system, which has been in place since the building opened, and is integrated with the broader enterprise asset management (EAM) system.

Completion Date: 1999
Height: 421 m (1,380 ft)
Stories: 88
Area: 289,500 sq m (3,116,152 sq ft)
Use: Hotel/Office
Owner/Developer: China Jin Mao Group Co. Ltd
Architect: Skidmore, Owings & Merrill (design); East China Architectural Design & Research Institute (architect of record)
Structural Engineer: Skidmore, Owings & Merrill
MEP Engineer: Skidmore, Owings & Merrill
Main Contractor: Shanghai Jin Mao Contractor
Other Consultants: Alan G. Davenport Wind Engineering BLWTL (wind); Edgett Willams Consulting Group, Inc. (vertical transportation); Permasteelisa Group (façade)

Jin Mao Energy-Saving Renovations

Year	Project and **Method**	Annual Energy Savings
2005	Cooling tower operation and control system renovation; **change group of 3 controls into individual control**	14,869 KWh
2008	Office area corridor lighting; **replace T8 bulbs with T5 bulbs**	470,000 kWh
2008	Lamp replacement in garage on Level B1; **replace 36 watt lamps with 28 watt lamps**	8,152 kWh
2008	Floodlighting in the lobby; **replace two metal halide lamps (150 watt) with three lamps (30 watt)**	22,075.2 kWh
2008	West fire stairs and east fire stairs; **replace double tube lamps (36 watt) with single tube lamps**	40,366 kWh
2008	Vestibules of the east and west fire stairs in office area; **replace 200 72-watt lamps with 100 36-watt lamps**	94,608 kWh
2009	Garage on Level B2 & B3; **use T5-28 watt /3 watt LED for Exit signs**	56,374 kWh
2009	Canopy lights outside the tower; **change 25 watt incandescent lamp into 5 watt energy-saving lamps**	17,500 kWh
2010	Podium building escalator; **speed down when not using**	5,000 kWh
2010	Landscape lighting for sightseeing lift; **change 25 watt incandescent lamp into 5 watt energy-saving lamps**	10,368 kWh
2011	Decorative lighting for the lobby elevator hall; **change 13 watt energy-saving lamps into 3 watt LED luminaires**	35,179 kWh
2011	Lighting for elevator cabinet; **change 36 watt fluorescent lamps to 5.4 watt LED luminaires**	13,795 kWh
2012	Lighting for steel escape ladder of podium; **replace 95 sets of lamps and adjusted location for several lamps**	24,624 kWh
2012	Lighting for PH floor; **replace 107 sets of lamps, loop circuit modification and BA timing control**	113,832 kWh
2013	The elevator energy feedback device; **add elevator energy feedback devices for elevators in sector one**	21,000 kWh

Management tracks a variety of performance metrics, including electricity, water, and natural gas consumption from month to month, and maintains key performance indicators around non-energy metrics, such as preventative maintenance, fixed asset purchases, requests for repairs, complaints, cost analysis, and equipment information records. The paperless processing associated with the system also contributes to the building's low carbon-emission footprint. Beginning in August 2013, to promote indoor air quality, the building's managers began tracking PM 2.5 particulate values in office areas and broadcasting the results daily on social media.

These performance measurement approaches have been augmented by consulting with the Association of German Engineers, which has helped Jin Mao develop computerized equipment management systems to help maintenance staff optimize the equipment life cycle, quality, and cost. A fluid energy metering system measures the flow of water through 89 sensors

Opposite: Overall view of tower

Left: Table showing energy-saving upgrades implemented over the years to reduce the building's electrical load

Jury Statement

For over 10 years, the management team of the Jin Mao building has meticulously tracked energy (oil, gas, and electricity) and water use, and have planned and documented a series of upgrades and improvements. They are clearly focused on where and how the energy was used and their commitment to performance is commended. This project was an early adopter of many innovative measures to save energy, and, though it is now more than a decade old, continues to set the new standard for high-rise buildings in China.

distributed throughout the building. An electricity metering system remotely and automatically measures electricity consumption of large equipment through a network of 300 sensors, allowing comparison of current and historical energy consumption. Together, these systems provide an objective data foundation for energy-use analysis.

Day-to-day maintenance of the aging equipment has not been enough to keep pace with the building's efficiency and environmental goals. Each month, building managers hold an energy-consumption analysis meeting and clarify the energy expenditure of each main equipment category, including water, electricity, and natural gas. The management team actively compares year-on-year and month-to-month consumption statistics, referencing the building automation system's daily control log, to make informed decisions about where energy-saving strategies should be implemented. Impacted facilities and equipment have been upgraded and/or renovated accordingly.

Darling Quarter

Sydney, Australia

While the Darling Quarter project is not a particularly tall building, the Technical Jury wanted to recognize the methodology by which this building measured its performance and presented its findings, in the form of an easy-to-read Base Building Energy Tracking Report, as an example of how the tall building industry in general could improve its monitoring and reporting regimes. The Darling Quarter monitoring project is exemplary, and illustrates the value of monitoring, tracking, and transparency in the cause of continuous improvement.

Darling Quarter consists of a high-performance building envelope with double glazing units, external shading and motorized blinds on the west façade and on the atrium roof, minimizing glare and reducing solar heat gain. This is coupled to a highly efficient, 100-percent fresh air chilled beam system that provides cooling and heating to the office areas. Additional features include a tri-generation system that offsets greenhouse gas (GHG) emissions and a blackwater plant system that provides recycled water for toilet flushing and the cooling tower makeup water.

The performance has been measured and verified through 14 months of energy monitoring, by collecting data from the building management system and comparing against the energy model used to predict the design energy use. At the end of the monitoring process, performance was verified and the building achieved a formal NABERS Energy rating, which exceeded the 5 Star target set during the design stage.

The key performance metric is measured in terms of GHG emissions under the well-established NABERS rating used, in this case, to measure the base building performance. The energy monitoring process showed the building achieved significant improvement in carbon emissions reduction when compared to the majority of office building stock existent in Sydney. A typical office building in Sydney is estimated to be equivalent to 2.5 NABERS Stars in operation. Darling Quarter achieved a 65 percent energy consumption reduction when compared to a typical building, which equates to a savings of approximately 5.7 million kg CO_2 per year, during the monitoring and tuning period.

The monitoring process ensured the building achieved a 5.5 Star rating in 2012, with a 30 percent reduction below its target of 5 Stars. The last year of operation shows the building performing 15–20 percent below a 5 Star NABERS Energy. The owner continues to monitor the energy performance and GHG of the building yearly.

Completion Date: September 2011
Height: 42 m (138 ft)
Stories: 10
Area: 55,000 sq m (592,015 sq ft)
Use: Office
Owner/Developer: Lend Lease
Architect: Francis-Jones Morehen Thorp
Structural Engineer: Arup
MEP Engineer: Arup; Aurecon
Main Contractor: Lend Lease

Darling Quarter – Tracking of Progressive Base Building Normalized Carbon Emission with Trigen in Operation

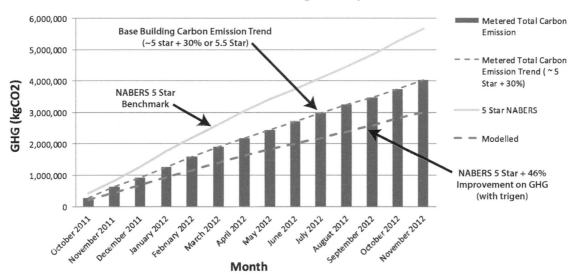

Top: Diagram comparing the GHG emissions of the building under operation against the energy model and 5 Star NABERS benchmark. The building operates at about 30 percent below the benchmark, but 30 percent higher than the energy model. The overage is likely attributed to longer hours of occupancy, additional retail equipment, and the tri-generation working less hours than expected at the time when the building was being fine-tuned

Left: Detail view of the western double glazing façade and motorized blinds

Right: Public space on the top floor of the atrium with shading devices protecting the glazed roof

CTBUH Lifetime Achievement

Awards Criteria

Lynn S. Beedle Award

The award recognizes an individual who has made extraordinary contributions to the advancement of tall buildings and the urban environment during his or her professional career. These contributions and leadership are recognized by the professional community and have significant effects, which extend beyond the professional community, to enhance cities and the lives of their inhabitants. The individual's contributions may be well known or little known by the public and may take any form, such as completed projects, research, technology, methods, ideas, or industry leadership.

The candidate may be from any area of specialization, including, but not limited to: architecture, structure, building systems, construction, academia, planning, development, or management. The award emphasizes the unique, multi-disciplinary nature of the Council and is thus set apart from other professional organizations' awards for single disciplines.

Fazlur R. Khan Medal

The award recognizes an individual for his/her demonstrated excellence in technical design and/or research that has made a significant contribution to a discipline(s) for the design of tall buildings and the built urban environment. The contribution may be demonstrated in the form of specific technical advances, innovations, design breakthroughs, building systems integration, or innovative engineering systems that resulted in a practical design solution and completion of a project(s). The consideration may be based on a single project or creative achievement through multiple projects.

In the case of both Lifetime Achievement Awards, the candidate may or may not be a member of the Council. The contributions of the award recipients should be generally consistent with the values and mission of the CTBUH and its founder, Dr. Lynn S. Beedle. The awards are not intended to be awarded posthumously, although they may be so awarded in some cases. The two Lifetime Achievement Awards are selected by the CTBUH Board of Trustees.

Winner
Lynn S. Beedle Lifetime Achievement Award

Douglas Durst
The Durst Organization

Opposite: Bank of America Tower, New York City, 2009 (366 m / 1,200 ft), CTBUH Best Tall Building Americas Award Winner (2010) and the first LEED Platinum Skyscraper

Above: Douglas Durst

"As the developer of the Conde Nast and Bank of America Towers, Douglas Durst has delivered on the promise of the sustainable tall building. While others only spoke, he took action."

David Malott, CTBUH Trustee, Kohn Pedersen Fox

One of the best ways to understand Douglas Durst's importance to the tall building industry, and the real estate field in general, is to look at his stewardship and advancement of a family company that has been investing in New York over 100 years of dramatic change. Douglas Durst is the chairman and a member of the third generation to lead the Durst Organization, which was founded in 1915 and is one of New York's oldest and most respected privately held owner-builder-managers of commercial and residential real estate. Its portfolio now comprises more than 13 million square meters of Class "A" Manhattan office space, as well as over 1,500 residential units. Today, The Durst Organization has two mixed-use residential rental buildings in development, with 1,200 units in the pipeline.

The Durst Organization began building on Third Avenue in the 1950s, and by the 1960s had helped establish the East Side as a commercial business district. In the late 1960s the company turned its attention to

Sixth Avenue, transforming it into Manhattan's premier corporate thoroughfare. Douglas Durst joined The Durst Organization in 1968, learning the business from his father, Seymour, and two uncles, Roy and David.

Rather than remain content to protect the family legacy, Durst struck out in several bold new directions. Far ahead of the current sustainability consciousness, in the 1980s, Durst installed energy-efficient light bulbs and variable-frequency fan drives in the company's existing portfolio of buildings.

In the mid-1990s, Times Square still had a seedy reputation. But Durst began developing 4 Times Square, lending the famous crossroads some overdue architectural sophistication. It was the first multi-tenanted project of its size to adopt standards for energy efficiency, sustainable materials, and indoor environmental quality, as well as for responsible construction, operations, and maintenance practices. It was also the first large-scale office tower built in New York after the real estate market collapse of the late 1980s. Housing Condé Nast publishing group and its famous Frank Gehry-designed cafeteria, the building contributed to Times Square's revival in the 2000s.

Since the completion of 4 Times Square, The Durst Organization has built two large-scale LEED Gold residential rental buildings – The Helena and The Epic. The 2005 Helena was the first voluntarily sustainable high-rise residential building to be constructed in New York City.

Currently under construction directly next to The Helena is the West 57th Street project. Designed by BIG-Bjarke Ingels Group, the building is a hybrid between the European perimeter block and a traditional Manhattan high-rise. Scheduled to open in 2015, Durst's latest residential tower is designed to allow the courtyard to have open views toward the Hudson River, bringing low western sun deep into the space.

Just a block away from 4 Times Square, in 2009, Durst added to his company's already impressive record along Sixth Avenue by creating a truly 21st-century building, The Bank of America Tower at One Bryant Park, recognized as the CTBUH 2010 Best Tall Building Americas. It was also the first high-rise office tower to be certified LEED Platinum by the US Green Building Council. One Bryant Park was designed to set a new standard in high-performance buildings, emphasizing the importance of occupant connections to nature and addressing the local environment. The site sits atop nearly a dozen subway lines and is within walking distance of three of the largest intermodal transportation hubs in North America.

At the height of confusion and malaise around the fate of the World Trade Center site in 2010, the Durst Organization under Douglas Durst assumed ownership, management, leasing, and operations responsibility for One World Trade Center, the tallest building in North America and a symbol of resilience after the 9/11 attacks. When Durst purchased the building, it had only

one tenant committed, which planned to occupy less than 10 percent of its 288,000 square meters.

"Despite these risks, we believe that New York and Lower Manhattan is a great bet, and the benefits of new and sustainable construction provide a critical edge," Durst said. As of this writing, the building is more than 56 percent leased.

Insight into the motivation for taking these risks can be found in some memorable statements Durst has made about his investing philosophy over the years.

"My experience is almost completely New York-centric," Durst said. "My grandfather and father were in real estate and my father had a strict policy of not buying anything that wasn't within walking distance of his house. I had the good fortune that he lived in Midtown Manhattan. Real Estate is always local. I am very lucky to work in one of the most dynamic and challenging real estate cities. My dad said, 'to build in New York, you need an architect, an engineer and two psychiatrists.' Today, you need two architects, two engineers and six psychiatrists. The risk, competition and regulation are intense, but so is the reward."

"Wherever you invest, it is important to remember that real estate is a service industry, not a commodity," Durst also said. "We plan for our children and grandchildren, not for the next earnings report. We build more efficient buildings, not only because they use less energy, are less expensive to operate, and

provide a more productive work environment, but because we are focused on providing not just an economic future for our children, but a healthy one as well."

Douglas was born in New York City in 1944 and graduated from the Fieldston School and the University of California Berkeley. Durst is a director of the Real Estate Board of New York, The New School, The Trust for Public Land, and Project for Public Spaces. Durst has been involved with the theatrical arts for many years and is a member of the board of directors of The Roundabout Theater and Primary Stages. Along with other family members, he is a trustee of The Old York Foundation, established by his father, which is committed to helping people through education to understand the history and issues facing New York City. In addition, Durst has been an environmental activist for many years and created one of the largest organic farms in New York State.

"There are few developers in the world who consistently push beyond the status quo in high risk cities to innovate and raise the bar in tall buildings. The Durst Organization led by Douglas Durst is an exception. Douglas has conceived and delivered buildings we can all learn from and encourages the industry to strive for achieving higher quality, sustainability, and performance."

Timothy Johnson, CTBUH Chairman, NBBJ

Winner
Fazlur R. Khan Lifetime Achievement Medal

Peter Irwin
Rowan Williams Davies & Irwin Inc.

Opposite: Peter Irwin stands with the 1:50 scale model of the top section of Burj Khalifa

Above: Peter Irwin

"Peter's deep understanding of wind engineering and his innovative thinking has added greatly to the design of many of today's tallest towers."

William Baker, CTBUH Trustee, SOM

Peter Irwin, one of the founding partners of Rowan Williams Davies and Irwin Inc. (RWDI), has led wind engineering on many of the world's tallest buildings, including the Petronas Towers, Taipei 101, Burj Khalifa, the Trump International Hotel & Tower Chicago, and Shanghai Tower. He draws on the fundamental principles of aerodynamics and structural response, using the wind tunnel and special analysis methods to study the effects of building shape, mass, stiffness, and damping. These principles are important, but Irwin also believes strongly in blending them with the dictates of architectural aesthetics and structural design. Throughout his career, Irwin has always argued that successful tall buildings require close collaboration between the architect, developer, structural engineer, and wind engineer.

The principle of collaboration permeates all aspects of the way Irwin's firm, RWDI, runs projects. The firm's motto, "complex issues made simple," encourages the idea that, even though many of the specialized

techniques of wind engineering are complex, in the end the results must be expressed in simple terms that can readily inform decision-making by the design team as a whole. An example of the collaborative approach is the one-day workshop frequently held at RWDI's laboratories, where the architect, developer, structural engineer, and wind engineering team are all present. Critical results are viewed online as they come out of the wind tunnel, and effects of changes to the design can be rapidly evaluated in terms of aerodynamic performance, aesthetics, functionality, cost, and structural feasibility.

In the 1980s, Irwin worked closely with structural engineer Jacob Grossman on several New York residential buildings of unprecedented slenderness (a greater than 10:1 height-to-width ratio), including the Metropolitan Tower and the City Spire. These towers pushed the limits of current motion criteria in the upper floors, and the experience gained in the design, testing, and subsequent performance of those buildings helped to set new standards for occupant comfort now used on many buildings. Irwin has also been instrumental in development of criteria for the comfort of pedestrians in the winds around tall buildings. The Irwin Sensor that he developed for wind tunnel studies of pedestrian impact is now widely used at many laboratories around the world.

In the late 1990s, Irwin was working with structural engineer Chris Stefanos on the Park Hyatt tower in Chicago, when wind studies showed that the building's structure should not be built without supplemental damping. As a result, the final design incorporated a 300-ton pendulum tuned mass damper (TMD) designed by RWDI's subsidiary firm Motioneering Inc. This was the first use of a pendulum TMD in a tall building in North America, and it precipitated an increase in use of damping systems on the continent, after a hiatus of almost 20 years. RWDI/Motioneering has since been responsible for the design and commissioning of more than 20 supplemental damping systems in North America and around the world, including the 700-ton spherical TMD in Taipei 101, a 600-ton TMD in the Trump World Tower in New York, a 600-ton Tuned Liquid Damper (TLD) in the Wall Center in Vancouver, and a 1300-ton TLD in the Comcast Center in Philadelphia.

Irwin adheres to the principle that on each project something can be learned, even those that are not built. The wind engineering and structural design of Cesar Pelli's Petronas Towers in Kuala Lumpur, represented an evolution from the plans for the unbuilt 610-meter Miglin Beitler Tower, Chicago, which was extensively studied by Irwin and the structural engineering team at Thornton Tomasetti, led by Charles Thornton. SOM's 828-meter Burj Khalifa tower in Dubai, which would exceed the previous world height record by 61 percent, evolved from previous studies by SOM and RWDI for the unbuilt Dearborn Tower in Chicago. For Burj Khalifa, Irwin initiated the first use of meso-scale

Opposite: Petronas Towers, Kuala Lumpur, 1998 (452 m / 1,483 ft), and wind tunnel testing model (below)

Right: Trump World Tower, New York City, 2001 (262 m / 861 ft), and the wind testing model (below)

249

meteorological modeling to evaluate wind behavior at the tops of extremely tall structures. This approach was again applied by RWDI in its recent studies of the Kingdom Tower in Saudi Arabia.

Born in the UK, Irwin received his bachelor's and master's degrees in aeronautical engineering from Southampton University. After two years of research at the Royal Aircraft Establishment in Farnborough, in 1974 he completed his PhD in mechanical engineering at McGill University, Montreal, Canada, and commenced his career in wind engineering at the National Research Council of Canada in Ottawa. In 1980, he joined MHTR, a small wind and snow consulting firm, and in 1986 was one of the founding partners of RWDI, which has since grown to be the largest wind engineering firm worldwide. He was president of RWDI from 1999 to 2008, during which time the firm's business volume grew 300 percent, and was chairman from 2008 to 2012.

Irwin is a Fellow of the American Society of Civil Engineers, The Structural Engineering Institute, the Canadian Society of Civil Engineers, The Engineering Institute of Canada, and the Canadian Academy of Engineering. He has also been the recipient of a number of awards, including the American Association of Wind Engineering's Industry Innovation Award (2013), ASCE's Jack E. Cermak Medal (2007), the Coopers Hill Medal (2003) of the Institute of Civil Engineers, UK, and the Canadian Society of Civil Engineers'

Gzowski Medal (1995). He was featured as a leader in structural engineering in Civil and Structural Engineer Magazine's January 2014 issue.

He has served on the CTBUH Board of Trustees from 2009 to 2012, and continues to be co-chair of the Wind Engineering Working Group. Irwin is one of the principal authors of the CTBUH Technical Guide *Wind Tunnel Testing of High-Rise Buildings*. He has also chaired and spoken at numerous CTBUH conferences, including New York, Dubai, and Mumbai.

He has contributed significantly to the development of codes and standards worldwide through participation in the Standing Committee on Structural Design for the Canadian Building Code and the wind committees of the ASCE 7 and ISO standards.

CTBUH 2014 Fellows

CTBUH Fellows are recognized for their contribution to the Council over an extended period of time, and in recognition of their work and the sharing of their knowledge in the design and construction of tall buildings and the urban habitat.

Johannes de Jong
KONE Corporation, Finland

Johannes de Jong joined KONE after receiving his master's degree in Engineering from the Polytechnical University of Delft in the Netherlands in 1977, and has risen through various research and development roles to become KONE's Head of Technology. De Jong is widely recognized in the industry as a vertical transportation expert, and has published several books and research papers on the subject. He has been involved in many of the world's tallest buildings. Through his research and development work, he holds more than 500 patents.

De Jong's involvement in CTBUH has been exemplary, and his enthusiasm and energy have become well known within the organization. He has been a member of the CTBUH Advisory Group since 2008, its Competitions Committee in 2012, and has published multiple papers in CTBUH publications. He has spoken at six CTBUH conferences since 2001.

Peter Weismantle
Adrian Smith + Gordon Gill Architecture, USA

Peter Weismantle is responsible for overseeing the technical development of Adrian Smith + Gordon Gill Architecture's supertall projects from concept to completion. He has developed widely recognized expertise in technical aspects of the exterior wall, façade access and vertical transportation in supertall buildings. His portfolio includes the Burj Khalifa, Dubai; Trump Tower, Chicago; Jin Mao Tower, Shanghai; and the under-construction Kingdom Tower in Jeddah.

Weismantle has been Chair of the CTBUH Height Committee since 2007, overseeing multiple changes to the CTBUH's height criteria. He presided over the Height Committee's 2013 ratification of the height of One World Trade Center as being officially recognized as 1,776 feet tall.. He is also chair of the CTBUH Advisory Group, and has been part of the peer review panel for the CTBUH International Research Seed Funding program, and is currently co-chair of the Façade Access Working Group.

CTBUH 2014 Awards Jury

The Awards Jury is appointed annually to determine the CTBUH Award Winners and Finalists. The multidisciplinary Main Jury determines the regional Best Tall Buildings, 10 Year Award, and Urban Habitat Award winners and finalists, while the Technical Jury, representing various engineering disciplines, selects the Innovation and Performance Award winners and finalists, as well as providing technical insights to the Main Jury on the best tall building awards.

Main Jury

Jeanne Gang *2014 Awards Jury Chair*
Studio Gang Architects | Chicago, USA

Jeanne Gang is the founder and principal of Studio Gang Architects, whose inventive tall buildings engage site, culture, and people. Studio Gang's Aqua Tower brings sustainability and a sense of community to high-rise living and has become a much lauded addition to the Chicago skyline.

Sir Terry Farrell
Farrells | London, UK

Sir Terry Farrell is the founding partner of Farrells. In 2013, he was voted the individual who made the greatest contribution to London's planning and development over the last 10 years and he recently published a national review of architecture and the built environment at the invitation of the UK Culture Minister.

David Gianotten
OMA | Hong Kong, China

David launched the OMA Hong Kong office in 2009, and became partner in 2010. David leads OMA's large portfolio in the Asia Pacific region, including the end stages of the CCTV tower in Beijing, and the design and execution of the Shenzhen Stock Exchange Tower and Taipei Performing Arts Centre.

Saskia Sassen
Columbia University | New York, USA

Saskia Sassen is the Robert S. Lynd Professor of Sociology and Co-Chair, The Committee on Global Thought at Columbia University. She has received diverse awards, from multiple doctor honoris causa to being chosen as one of the Top 100 Global Thinkers by Foreign Policy-2011.

Wai Ming (Thomas) Tsang
Shenzhen Ping An Financial Centre Construction and Development Ltd. Co. | Hong Kong, China

Thomas joined Ping An Real Estate to assume the position as the CEO of Shenzhen Ping An Financial Centre Construction and Development Limited Company in April 2012. His responsibility is to oversee the construction and Development of the PAFC project.

Antony Wood
CTBUH / IIT / Tongji University | Chicago, USA

Antony Wood is the Executive Director of the CTBUH, responsible for the day-to-day running of the Council. Antony is also a Research Professor in the College of Architecture at the Illinois Institute of Technology, and a Visiting Professor of Tall Buildings at Tongji University, Shanghai.

Technical Jury

David Scott *2014 Technical Jury Chair*
Laing O'Rourke | London, UK

David is the Lead Structural Director at Laing O'Rourke, and is a past Chairman of the CTBUH (2006–2009). David has a passion for unusual structures and has extensive tall building experience. He is an advocate for low-energy buildings and off-site construction.

Guo-Qiang Li
Tongji University | Shanghai, China

Guo-Qiang Li is a Professor of Structural Engineering at the College of Civil Engineering in Tongji University, the director of the National Research Center for Industrialized Construction of China, and the director of the Research Center of the Education Ministry of China for Steel Construction.

Nengjun Luo
CITIC HEYE Investment Co., Ltd. | Beijing, China

As Deputy General Manager of CITIC HEYE Investment Co., Ltd., Dr. Luo is responsible for the overall planning and operation control of project management, design, construction of the CITIC Headquarters building (528 meters), which will become the tallest building in Beijing upon completion.

Douglas Mass
Cosentini Associates | New York, USA

Douglas is the President of Cosentini Associates. He is internationally recognized in the design and construction industry because of his expertise as a mechanical engineer, his commitment to design innovation, and a long list of high-performance building projects designed by leading architects.

Paul Sloman
Arup | Sydney, Australia

Paul Sloman is a Principal and the Buildings Group Leader at Arup Sydney. With over 20 years experience in the design of major flagship buildings and mixed developments all over the world, Paul brings to his projects a wealth of international experience and global connectivity.

Peter Williams
AECOM | London, UK

As the leader of AECOM's Building Engineering practice in EMEA, Peter balances his activities in the UK office sector with a broad spectrum of building projects internationally. He works with multidiscipline structures, services, and specialist design teams.

Review of Last Year's CTBUH 2013 Awards

November 7, 2013 marked the 12th annual CTBUH Awards Symposium, Ceremony & Dinner, held in the Illinois Institute of Technology's S.R. Crown Hall.

The tone of the Symposium was set definitively early on by Henry Cobb, the 2013 Lynn S. Beedle Lifetime Achievement Award recipient, who argued that height is far from the sole determinant of good design in tall buildings. The founding partner of Pei Cobb Freed & Partners has been practicing architecture with a near-legendary level of integrity for 56 years. This was his second appearance on the Symposium stage in as many years, having collected the Best Tall Building Europe award for Palazzo Lombardia in 2012.

"I'm interested in sustainability – not just environmental sustainability but social sustainability," Cobb said. "For a tall building, social sustainability has more to do with the way it greets the ground than the way it reaches the sky."

Through a career spanning 59 years, Clyde N. Baker, Jr., Senior Principal Engineer (Retired), AECOM, winner of the 2013 Fazlur R. Khan Lifetime Achievement Medal, has become synonymous with geotechnical engineering for tall buildings, devoting passionate energy and innovation to the practice. While accepting the awards he joked, "my architect and structural friends probably don't agree that *I* designed all those buildings…", though they certainly wouldn't disagree he played a critical role.

In 2013, the jury simply could not choose between two excellent candidates for the CTBUH Innovation Award, and thus awarded co-winners. Shuguang Wang, Executive Manager, Broad Group, presented the Broad Sustainable Building Prefabricated Construction Method. The effectiveness of this method was exemplified by a demonstration of the execution of the concept through the T30 building in Changsha, China, which gained notoriety through a time-lapse YouTube video that showed the building being constructed in 15 days.

Wang described the interior of T30, including the equipment that had been designed for it, such as four-layer windows, automatic curtains, LED lighting, dynamic air supply and thermal insulation. In all, the Broad Group will have built more than 30 such buildings by the end of 2013, Wang said.

The co-winner of the 2013 Innovation Award was KONE's carbon-fiber UltraRope™ for elevator lifting. Johannes de Jong, Head of Technology at KONE, explained the challenges of current high-rise elevator technology, and positioned the genesis of UltraRope as a solution to issues of energy consumption in the elevator industry.

UltraRope's carbon-fiber composition reduces its weight against that of a steel rope to the degree that using UltraRope would result in 60 percent less energy than using steel rope in a comparable single lift, de Jong said.

Opposite Top: Henry Cobb, Pei Cobb Freed & Partners, winner of the Lynn S. Beedle Award, presents during the Awards Symposium

Opposite Bottom: Clyde Baker, AECOM (retired), winner of the Fazlur R. Khan Medal, accepts the awards from Tim Johnson, CTBUH Chairman/NBBJ Design Partner

Top: Over 500 delegates fill the auditorium to hear presentations from the 2013 winners

Middle Left: Shuguang Wang presents the Broad Sustainable Building Prefabricated Construction Process, 2013 Innovation Award co-winner

Bottom Left: Johannes de Jong, KONE, presents UltraRope, 2013 Innovation Award co-winner

The 2013 Awards also introduced the new CTBUH 10 Year Award which recognizes proven value and performance of a completed building, across one or more of a wide range of criteria. The uniquely shaped Gherkin was the inaugural winner of this award, and was collected by Jimmy Demetriou of CBRE 30 St Mary Axe Management Ltd. Jimmy recognized the team effort behind the landmark project having delivered "the vision of a true tower for London, which will hopefully continue to inspire generations for years to come."

Top Left: 30 St Mary Axe, London, winner of the inaugural CTBUH 10 Year Award

Bottom Left: 2013 Awards Jury Chair, Jeanne Gang, Studio Gang Architects, presents the 10 Year Award to Jimmy Demetriou, CBRE 30 St Mary Axe Management Ltd.

Top Right: The Bow, Calgary, winner of the 2013 Best Tall Building Americas award

Bottom Right: The Bow team (left to right: Nigel Dancey, Foster + Partners; Ken Hornby, Encana; Jack Matthews, Matthews Southwest; and James Barnes, Foster + Partners) pose for a photo after accepting the Best Tall Building Americas award

Opposite Top: The Shard, London, winner of the Best Tall Building Europe award

Opposite Bottom: Dave Elder, Mace, and Bill Price, WSP Group, accept the Best Tall Building Europe award from 2013 Awards Jury Chair, Jeanne Gang, Studio Gang Architects

The Bow, winner of the Best Tall Buildings Americas award, is both stunning as a form and functions well from an environmental and urban standpoint, especially in the context of a harsh northern climate. It serves as a rare example of an iconic design resulting from the most practical, yet creative, response to site constraints. The resolution of wind loading, light access, thermal comfort, and public space objectives has resulted in a solution that embodies synthesis but bears no hint of compromise.

The project was presented at the symposium by developer Jack Matthews, President of Matthews Southwest and Nigel Dancey, of Foster + Partners.

Dancey described the Bow as being a part of a long tradition of high-performing, contextual buildings at Foster + Partners, while improving upon some of the more adventurous aspects of prior buildings, such as 30 St Mary Axe, London, which used the diagrid exoskeleton structural system, and Commerzbank Headquarters, Frankfurt, which was an early experiment in sky gardens. The braced-frame-and-diagrid approach used 20 percent less steel by weight than a conventional structure, Dancey said.

The developers of The Shard, winner of the Best Tall Building Europe award, showed remarkable tenacity in bringing it to fruition. The level of determination to wring economic success and poetics out of the project while still supporting public life at street level is remarkable. Through more than a decade of design revisions and public inquiries, the project team was unwavering in its determination to do more than impose a tall building on a neglected but architecturally rich neighborhood. Their determination was to secure the future of the London Bridge Quarter district itself.

Bill Price of WSP and Dave Elder of Mace presented the Shard. Even if the Shard did not single-handedly save the London property market, it remains a model of persistence. Two years of public inquiry and several more of funding uncertainty were only the beginning.

Elder explained the extraordinary feats of engineering that were employed on an extremely constrained site. The slab was poured continuously in 32 hours, and had less than 6 thousandth of an inch of shrinkage. In order to reduce the chance of weather delays, the team redesigned the original structural schedule from 1,200 pieces of steel in an 18-week program to 200 pieces to be installed over eight weeks.

No less of a commitment was on display during Jim Goettsch's presentation of Sowwah Square, winner

of the Best Tall Building Middle East & Africa award. While other teams grappled with context, Goettsch's team wrestled with the relative lack of it – Sowwah Square was built on a master-planned island off of Abu Dhabi that had been wiped mostly clean of all context.

Goettsch seemed bemused that the relatively low-slung buildings were selected by the jury, especially as they came from a region that contains the world's current and future tallest building. But it was the regional response that set the buildings apart from their peers.

"In the Middle East, if you are outside for six months of the year, the first thing you want to do is get into the shade – these buildings provide that," Goettsch said. "The sun rises very quickly, so each façade gets a completely different exposure. Each one warrants a different strategy."

Thus, the south façade used fixed shades while the east and west side had operable ones. The building also dealt with the harsh, humid climate outside, using

a double-skin façade to keep indoor temperatures at 23.8°C while outside it could be as warm as 60°C. The cavity in the middle mediates these temperatures, such that the inside and outside of the inner glass only vary in temperature by a few degrees, reducing the load on cooling systems, Goettsch explained. The total energy savings were about 28 percent against a conventional design.

The awards evening culminated in the announcement of the Best Tall Building Worldwide at the Ceremony and Dinner. Following the Symposium presentations, the Awards Jury convened to consider the four regional winners, ultimately selecting the CCTV Headquarters, Beijing (the 2013 Best Tall Building Asia & Australasia) as the overall Worldwide winner. CCTV's award was announced by Wiel Arets, Dean of the School of Architecture at IIT.

Rem Koolhaas, Founding Partner, Office for Metropolitan Architecture, delivered the winning presentation, entitled "A New Typology for the Skyscraper: CCTV Headquarters, Beijing." Koolhaas detailed the thinking behind CCTV Headquarters, Beijing, one of the most unusually shaped skyscrapers yet constructed. The knowledge that a master plan called for surrounding the building with 300 skyscrapers led OMA to rethink how to produce a sufficiently iconic building.

"It didn't make sense to introduce another needle or try to compete with height," Koolhaas said. "What was necessary was an entirely different language and condition of a skyscraper."

Opposite Left: Sowwah Square, Abu Dhabi, winner of the 2013 Best Tall Building Middle East & Africa

Opposite Right: James Goettsch, Goettsch Partners (fourth from the left), and the Sowwah Square team pose for a photo with the CTBUH 2013 Awards skyline, after accepting the Best Tall Building Middle East & Africa award

Top: CCTV, Beijing, winner of the Best Tall Building Asia & Australasia and overall Best Tall Building Worldwide

Bottom: Rem Koolhaas (center), Office for Metropolitan Architecture, and the CCTV team accept the Best Tall Building Asia & Australasia and Best Tall Building Worldwide trophies from IIT Dean of Architecture Wiel Arets (second from left)

Instead, the designers conceived a twisting loop that turns corners and appears to nearly fall back on itself from some angles. The loop takes occupants through more than 200 different functional spaces for an organization that is anything but monolithic, but is more like a city of interdependent broadcast-related entities. The amount of precision and bespoke design in CCTV was unprecedented. The tolerances on the two halves of the skybridge were so tight, the connection could only be fused at 6:00 in the morning, when the steel on both sides was sufficiently cool as to warrant against uncontrolled expansion.

"CCTV is perhaps a building that the Chinese never could have invented, but it's a building Europeans never could have built," Koolhaas said.

Upon receiving the 2013 Best Tall Building Worldwide award at the Awards Ceremony, Koolhaas said, "When I published my last book, *Content*, in 2003, one chapter was called 'Kill the Skyscraper.' Basically it was an expression of disappointment at the way the skyscraper typology was used and applied. I didn't think there was a lot of creative life left in skyscrapers. Therefore, I tried to launch a campaign against the skyscraper in its more uninspired form.

"The fact that I am standing on this stage now, in this position, meant that my declaration of war went completely unnoted, and that my campaign was completely unsuccessful," Koolhaas joked, concluding, "Being here, it is quite moving – to be part of a community that is trying to make skyscrapers more interesting. I am deeply grateful, and thank all my partners."

The audience vote, taken separately, submitted via text message, and kept from the jury's view until after their verdict had been announced, was the same – CCTV was the winner of the popular vote, too.

Overview of All Past Winners

Best Tall Building – Overall & Best Tall Building – Asia & Australasia
CCTV Headquarters
Beijing, China

Best Tall Building – Americas
The Bow
Calgary, Canada

Best Tall Building – Europe
The Shard
London, UK

Best Tall Building – Middle East & Africa
Sowwah Square
Abu Dhabi, UAE

10 Year Award Winner
30 St Mary Axe
London, UK

Tall Building Innovation Award Co-Winner
Broad Sustainable Building Prefabricated Construction Process

Tall Building Innovation Award Co-Winner
KONE UltraRope™

Best Tall Building – Overall & Best Tall Building – Middle East & Africa
Doha Tower,
Doha, Qatar

Best Tall Building – Americas
Absolute Towers,
Mississauga, Canada

Best Tall Building – Asia & Australasia
1 Bligh Street, *Sydney, Australia*

Best Tall Building – Europe
Palazzo Lombardia,
Milan, Italy

Tall Building Innovation Award
Al Bahar Towers,
Abu Dhabi, UAE

Best Tall Building – Overall & Best Tall Building – Europe
KfW Westerkade,
Frankfurt, Germany

Best Tall Building – Americas
Eight Spruce Street,
New York City, USA

Best Tall Building – Asia & Australasia
Guangzhou International Finance Center,
Guangzhou, China

**Best Tall Building –
Middle East & Africa**
The Index,
Dubai, UAE

**"Global Icon" Award
& Best Tall Building –
Middle East & Africa**
Burj Khalifa,
Dubai, UAE

**Best Tall Building – Overall
& Best Tall Building –
Europe**
Broadcasting Place,
Leeds, UK

**Best Tall Building –
Americas**
Bank of America Tower,
New York City, USA

**Best Tall Building –
Asia & Australasia**
Pinnacle @ Duxton,
Singapore

**Best Tall Building – Overall
& Best Tall Building –
Asia & Australasia**
Linked Hybrid,
Beijing, China

**Best Tall Building –
Americas**
Manitoba Hydro Place,
Winnipeg, Canada

Best Tall Building – Europe
The Broadgate Tower,
London, UK

**Best Tall Building –
Middle East & Africa**
Tornado Tower,
Doha, Qatar

**Best Tall Building – Overall
& Best Tall Building –
Asia & Australasia**
Shanghai WFC,
Shanghai, China

**Best Tall Building –
Americas**
The New York Times Building,
New York City, USA

Best Tall Building – Europe
51 Lime Street,
London, UK

**Best Tall Building –
Middle East & Africa**
Bahrain World Trade Center,
Manama, Bahrain

Best Tall Building – Overall
Beetham Tower,
Manchester, UK

**Best Sustainable Tall
Building**
Hearst Tower,
New York City, USA

Overview of All Past Winners (continued)

2013

Lynn S. Beedle Award
Henry Cobb

Fazlur R. Khan Medal
Clyde N. Baker, Jr.

2012

Lynn S. Beedle Award
Helmut Jahn

Fazlur R. Khan Medal
Charles Thornton

Fazlur R. Khan Medal
Richard Tomasetti

2011

Lynn S. Beedle Award
Adrian Smith

Fazlur R. Khan Medal
Dr. Akira Wada

2010

Lynn S. Beedle Award
William Pedersen

Fazlur R. Khan Medal
Ysrael A. Seinuk

2009

Lynn S. Beedle Award
John C. Portman, Jr.

Fazlur R. Khan Medal
Dr. Prabodh V. Banavalkar

2008

Lynn S. Beedle Award
Cesar Pelli

Fazlur R. Khan Medal
William F. Baker

Lynn S. Beedle Award
Lord Norman Foster

Fazlur R. Khan Medal
Dr. Farzad Naeim

Lynn S. Beedle Award
Dr. Ken Yeang

Fazlur R. Khan Medal
Srinivasa "Hal" Iyengar

Lynn S. Beedle Award
Dr. Alan Davenport

Fazlur R. Khan Medal
Dr. Werner Sobek

Lynn S. Beedle Award
Gerald Hines

Fazlur R. Khan Medal
Leslie E. Robertson

Lynn S. Beedle Award
Charles DeBenedittis

Lynn S. Beedle Award
Dr. Lynn S. Beedle

CTBUH Height Criteria

The Council on Tall Buildings and Urban Habitat is the official arbiter of the criteria upon which tall building height is measured, and the title of "The World's (or Country's, or City's) Tallest Building" determined. The Council maintains an extensive set of definitions and criteria for measuring and classifying tall buildings which are the basis for the official "100 Tallest Buildings in the World" list.

What is a Tall Building?

There is no absolute definition of what constitutes a "tall building." It is a building that exhibits some element of "tallness" in one or more of the following categories:

a) Height relative to context:

It is not just about height, but about the context in which it exists. Thus whereas a 14-story building may not be considered a tall building in a high-rise city such as Chicago or Hong Kong, in a provincial European city or a suburb this may be distinctly taller than the urban norm.

b) Proportion:

Again, a tall building is not just about height but also about proportion. There are numerous buildings which are not particularly high, but are slender enough to give the appearance of a tall building, especially against low urban backgrounds. Conversely, there are numerous big/large footprint buildings which are quite tall but their size/floor area rules them out as being classed as a tall building.

c) Tall Building Technologies:

If a building contains technologies which may be attributed as being a product of "tall" (e.g., specific vertical transport technologies, structural wind bracing as a product of height, etc.), then this building can be classed as a tall building.

Although number of floors is a poor indicator of defining a tall building due to the changing floor-to-floor height between differing buildings and functions (e.g., office versus residential usage), a building of perhaps 14 or more stories – or over 50 meters (165 feet) in height – could perhaps be used as a threshold for considering it a "tall building."

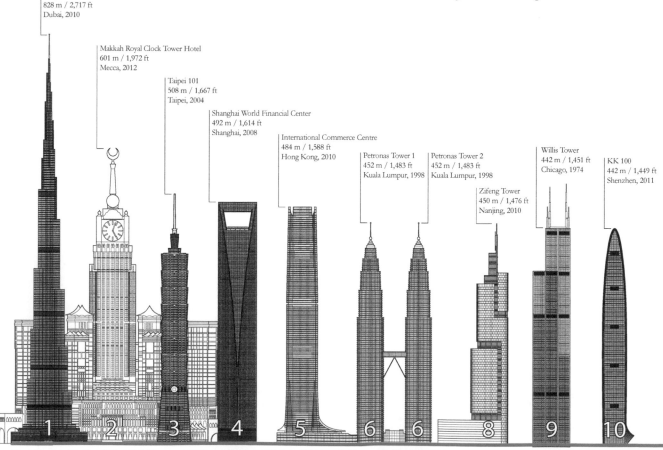

Burj Khalifa
828 m / 2,717 ft
Dubai, 2010

Makkah Royal Clock Tower Hotel
601 m / 1,972 ft
Mecca, 2012

Taipei 101
508 m / 1,667 ft
Taipei, 2004

Shanghai World Financial Center
492 m / 1,614 ft
Shanghai, 2008

International Commerce Centre
484 m / 1,588 ft
Hong Kong, 2010

Petronas Tower 1
452 m / 1,483 ft
Kuala Lumpur, 1998

Petronas Tower 2
452 m / 1,483 ft
Kuala Lumpur, 1998

Zifeng Tower
450 m / 1,476 ft
Nanjing, 2010

Willis Tower
442 m / 1,451 ft
Chicago, 1974

KK 100
442 m / 1,449 ft
Shenzhen, 2011

Above: Diagram of the World's Tallest 20 Buildings according to the CTBUH Height Criteria of "Height to Architectural Top" (as of May 2014)

What are Supertall and Megatall Buildings?

The CTBUH defines "supertall" as a building over 300 meters (984 feet) in height, and a "megatall" as a building over 600 meters (1,968 feet) in height. As of July 2014 there are 81 supertall and 2 megatall building completed and occupied globally. There are also currently an additional 100 supertalls and 5 megatalls under construction globally.

How is a Tall Building Measured?

The CTBUH recognizes tall building height in three categories:

1. Height to Architectural Top: Height is measured from the level[1] of the lowest, significant,[2] open-air,[3] pedestrian[4] entrance to the architectural top of the building, including spires, but not including antennae, signage, flagpoles or other functional-technical equipment.[5] This measurement is the most widely utilized and is employed to define the Council on Tall Buildings and Urban Habitat (CTBUH) rankings of the "World's Tallest Buildings."

2. Highest Occupied Floor: Height is measured from the level[1] of the lowest, significant,[2] open-air,[3] pedestrian[4] entrance to the finished floor level of the highest occupied[6] floor within the building.

3. Height to Tip: Height is measured from the level[1] of the lowest, significant,[2] open-air,[3] pedestrian[4] entrance to the highest point of the building, irrespective of material or function of the highest element (i.e., including antennae, flagpoles, signage, and other functional technical equipment).

Number of Floors:

The number of floors should include the ground floor level and be the number of main floors above ground, including any significant mezzanine floors and major mechanical plant floors. Mechanical mezzanines should not be included if they have a significantly smaller floor area than the major floors below. Similarly, mechanical penthouses or plant rooms protruding above the general roof area should not be counted. Note: CTBUH floor counts may differ from published accounts, as it is common in some regions of the world for certain floor levels not to be included (e.g., the level 4, 14, 24, etc. in Hong Kong).

Building Usage:

What is the difference between a tall building and a telecommunications/observation tower?

A tall "building" can be classed as such (as opposed to a telecommunications/observation tower) and is eligible for the "Tallest" lists if at least 50 percent of its height is occupied by usable floor area.

Single-Function and Mixed-Use Buildings:

A **single-function** tall building is defined as one where 85 percent or more of its total floor area is dedicated to a single usage.

A **mixed-use** tall building contains two or more functions (or uses), where each of the functions occupy a significant proportion[7] of the tower's total space. Support areas such as car parks and mechanical plant space do not constitute mixed-use functions. Functions are denoted on CTBUH "tallest" lists in descending order, e.g., "hotel/office" indicates hotel above office function.

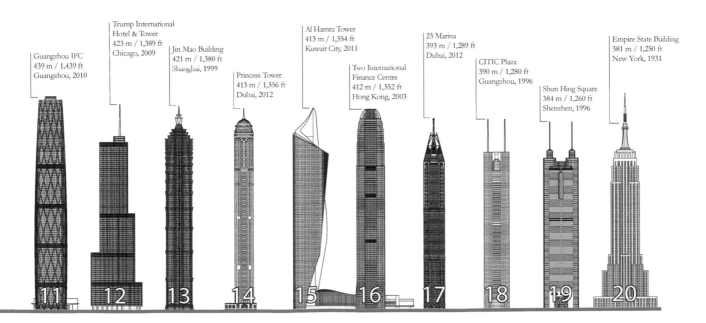

Guangzhou IFC
439 m / 1,439 ft
Guangzhou, 2010

Trump International Hotel & Tower
423 m / 1,389 ft
Chicago, 2009

Jin Mao Building
421 m / 1,380 ft
Shanghai, 1999

Princess Tower
413 m / 1,356 ft
Dubai, 2012

Al Hamra Tower
413 m / 1,354 ft
Kuwait City, 2011

Two International Finance Centre
412 m / 1,352 ft
Hong Kong, 2003

23 Marina
393 m / 1,289 ft
Dubai, 2012

CITIC Plaza
390 m / 1,280 ft
Guangzhou, 1996

Shun Hing Square
384 m / 1,260 ft
Shenzhen, 1996

Empire State Building
381 m / 1,250 ft
New York, 1931

11 12 13 14 15 16 17 18 19 20

Building Status:

Completed

A building is considered to be "Completed" (and officially added to the CTBUH Tallest Buildings lists) if it fulfills all of the following three criteria: (i) Topped out structurally and architecturally, (ii) Fully clad, (iii) Open for business, or at least partially occupiable.

Topped Out

A building is considered to be "topped out" when it is under construction, and has reached its full height both structurally and architecturally (e.g., including its spires, parapets, etc.)

Under Construction (Start of Construction)

A tall building is considered to be "under construction" when site clearing has been completed and foundation/piling work has begun.

On Hold

A building is considered to be "on hold" when construction works had begun, but work on-site has been halted indefinitely; however, there is still an intent to complete the construction to the original design at a future date.

Never Completed

A building is considered to be "never completed" when construction works had begun, but work on-site was halted and never resumed. The site may go on to accommodate a new building, different to the original design, that may or may not retain the original construction.

Proposed (Proposal)

A building is considered to be "Proposed" (i.e., a real proposal) when it fulfills all of the following criteria: (i) Has a specific site with ownership interests within the building development team, (ii) Has a full professional design team progressing the design beyond the conceptual stage, (iii) Has obtained, or is in the process of obtaining, formal planning consent/legal permission for construction, (iv) Has a full intention to progress the building to construction and completion. Only buildings that have been announced publicly by the client and fulfill all the above criteria are included in the CTBUH "proposed" building listings. The source of the announcement must also be credible. Due to the changing nature of early stage designs and client information restrictions, some height data may be unconfirmed.

Demolished

A building is considered to be "demolished" after it has been destroyed by controlled end-of-life demolition, fire, natural catastrophe, war, terrorist attack, or through other means intended or unintended.

Vision

A building is considered to be a "vision" when it is either: (i) In the early stages of inception and does not yet fulfill the criteria under the "proposal" category, or (ii) Was a proposal that never advanced to the construction stages, or (iii) Was a theoretical proposition.

Structural Material

A **steel** tall building is defined as one where the main vertical and lateral structural elements and floor systems are constructed from steel.

A **concrete** tall building is defined as one where the main vertical and lateral structural elements and floor systems are constructed from concrete.

A **composite** tall building utilizes a combination of both steel and concrete acting compositely in the main structural elements, thus including a steel building with a concrete core. A mixed-structure tall building is any building that utilizes distinct steel or concrete systems above or below each other. There are two main types of mixed structural systems: a steel/concrete tall building indicates a steel structural system located above a concrete structural system, with the opposite true of a concrete/steel building

Additional Notes on Structure:

(i) If a tall building is of steel construction with a floor system of concrete planks on steel beams, it is considered a steel tall building. (ii) If a tall building is of steel construction with a floor system of a concrete slab on steel beams, it is considered a steel tall building. (iii) If a tall building has steel columns plus a floor system of concrete beams, it is considered a composite tall building.

---

[1] Level: finished floor level at threshold of the lowest entrance door.

[2] Significant: the entrance should be predominantly above existing or pre-existing grade and permit access to one or more primary uses in the building via elevators, as opposed to ground floor retail or other uses which solely relate/connect to the immediately adjacent external environment. Thus entrances via below-grade sunken plazas or similar are not generally recognized. Also note that access to car park and/or ancillary/support areas are not considered significant entrances.

[3] Open-air: the entrance must be located directly off of an external space at that level that is open to air.

[4] Pedestrian: refers to common building users or occupants and is intended to exclude service, ancillary, or similar areas.

[5] Functional-technical equipment: this is intended to recognize that functional-technical equipment is subject to removal/addition/change as per prevalent technologies, as is often seen in tall buildings (e.g., antennae, signage, wind turbines, etc. are periodically added, shortened, lengthened, removed and/or replaced).

[6] Highest occupied floor: this is intended to recognize conditioned space which is designed to be safely and legally occupied by residents, workers or other building users on a consistent basis. It does not include service or mechanical areas which experience occasional maintenance access, etc.

[7] This "significant proportion" can be judged as 15 percent or greater of either: (1) the total floor area, or (2) the total building height, in terms of number of floors occupied for the function. However, care should be taken in the case of supertall towers. For example a 20-story hotel function as part of a 150-story tower does not comply with the 15 percent rule, though this would clearly constitute mixed-use.

100 Tallest Buildings in the World (as of July 2014)

The Council maintains the official list of the 100 Tallest Buildings in the World, which are ranked based on the height to architectural top, and includes not only completed buildings, but also buildings currently under construction. However, a building does not receive an official ranking number until it is completed.

Color Key:
Buildings in black are completed and officially ranked.
Buildings in blue are under construction and have topped out.
Buildings in green are under construction, but have not yet topped out.

Rank	Building Name	City	Height (meters)	Height (feet)	Stories	Year	Material	Function
	Kingdom Tower	Jeddah	1000+*	3,281+*	167	2019	concrete	residential / hotel / office
1	**Burj Khalifa**	**Dubai**	**828**	**2,717**	**163**	**2010**	**steel / concrete**	**office / residential / hotel**
	Suzhou Zhongnan Center	Suzhou	700+*	2,297+*	138	2020		residential / hotel / office
	Ping An Finance Center	Shenzhen	660	2,165	115	2016	composite	office
	Wuhan Greenland Center	Wuhan	636	2,087	125	2017	composite	hotel / residential / office
	Shanghai Tower	Shanghai	632	2,073	128	2015	composite	hotel / office
2	**Makkah Clock Royal Tower Hotel**	**Mecca**	**601**	**1,972**	**120**	**2012**	**steel / concrete**	**other / hotel / multiple**
	Goldin Finance 117	Tianjin	597	1,957	128	2016	composite	hotel / office
	Pearl of the North	Shenyang	565	1,854	111	2018		office
	Lotte World Tower	Seoul	555	1,819	123	2016	composite	hotel / office
	One World Trade Center	New York City	541	1,776	94	2014	composite	office
	CTF Finance Centre	Guangzhou	530	1,739	111	2016	composite	hotel / residential / office
	Tianjin Chow Tai Fook Binhai Center	Tianjin	530	1,739	97	2018	composite	residential / hotel / office
	Zhongguo Zun	Beijing	528	1,732	108	2018	composite	office
3	**Taipei 101**	**Taipei**	**508**	**1,667**	**101**	**2004**	**composite**	**office**
4	**Shanghai World Financial Center**	**Shanghai**	**492**	**1,614**	**101**	**2008**	**composite**	**hotel / office**
	Hengqin Headquarters Tower 2	Zhuhai	490	1,608	106	2017	composite	office
5	**International Commerce Centre**	**Hong Kong**	**484**	**1,588**	**108**	**2010**	**composite**	**hotel / office**
	Chongqing Corporate Avenue 1	Chongqing	468	1,535	100	2017	composite	hotel / office
	Guangdong Building	Tianjin	468	1,535	91	2017	composite	residential / hotel / office
	Lakhta Center	St Petersburg	463	1,517	86	2018	composite	office
	Riverview Plaza A1	Wuhan	460	1,509	82	2017		hotel / office
	Changsha IFS Tower T1	Changsha	452	1,483	88	2017	composite	residential / office
6	**Petronas Tower 1**	**Kuala Lumpur**	**452**	**1,483**	**88**	**1998**	**composite**	**office**
6	**Petronas Tower 2**	**Kuala Lumpur**	**452**	**1,483**	**88**	**1998**	**composite**	**office**
	Suzhou IFS	Suzhou	450	1,476	95	2017	composite	residential / hotel / office
8	**Zifeng Tower**	**Nanjing**	**450**	**1,476**	**66**	**2010**	**composite**	**hotel / office**
9	**Willis Tower**	**Chicago**	**442**	**1,451**	**108**	**1974**	**steel**	**office**
	World One	Mumbai	442	1,450	117	2015	composite	residential
10	**KK100**	**Shenzhen**	**442**	**1,449**	**100**	**2011**	**composite**	**hotel / office**
11	**Guangzhou International Finance Center**	**Guangzhou**	**439**	**1,439**	**103**	**2010**	**composite**	**hotel / office**
	Wuhan Center	Wuhan	438	1,437	88	2015	composite	hotel / residential / office
	106 Tower	Dubai	433	1,421	107	2018	concrete	residential
	Diamond Tower	Jeddah	432	1,417	93	2017		residential
	Dream Dubai Marina	Dubai	427	1,399	101	2015	concrete	serviced apartments / hotel
	432 Park Avenue	New York City	426	1,397	85	2015	concrete	residential
	225 West 57th Street	New York City	425+*	1,394+*	88	2018	concrete	residential / retail
12	**Trump International Hotel & Tower**	**Chicago**	**423**	**1,389**	**98**	**2009**	**concrete**	**residential / hotel**
13	**Jin Mao Tower**	**Shanghai**	**421**	**1,380**	**88**	**1999**	**composite**	**hotel / office**
14	**Princess Tower**	**Dubai**	**413**	**1,356**	**101**	**2012**	**steel / concrete**	**residential**

* estimated height

Rank	Building Name	City	Height (meters)	Height (feet)	Stories	Year	Material	Function
15	Al Hamra Tower	Kuwait City	413	1,354	80	2011	concrete	office
16	Two International Finance Centre	Hong Kong	412	1,352	88	2003	composite	office
	LCT Landmark Tower	Busan	412	1,350	101	2018		hotel / residential
	Huaguoyuan Tower 1	Guiyang	406	1,332	64	2017	composite	-
	Huaguoyuan Tower 2	Guiyang	406	1,332	64	2017	composite	-
	Nanjing Olympic Suning Tower	Nanjing	400	1,312	88	2017	steel / concrete	residential / hotel / office
	China Resources Headquarters	Shenzhen	400	1,312	70	2017		office
17	23 Marina	Dubai	393	1,289	90	2012	concrete	residential
18	CITIC Plaza	Guangzhou	390	1,280	80	1996	concrete	office
	Logan Century Center 1	Nanning	386	1,266	82	2017	composite	hotel / office
	Capital Market Authority Tower	Riyadh	385	1,263	79	2015	composite	office
19	Shun Hing Square	Shenzhen	384	1,260	69	1996	composite	office
	Eton Place Dalian Tower 1	Dalian	383	1,257	80	2014	composite	hotel / office
	Abu Dhabi Plaza	Astana	382	1,253	88	2017		residential
	World Trade Center Abu Dhabi - The Residences	Abu Dhabi	381	1,251	88	2014	concrete	residential
20	Empire State Building	New York City	381	1,250	102	1931	steel	office
21	Elite Residence	Dubai	380	1,248	87	2012	concrete	residential
22	Central Plaza	Hong Kong	374	1,227	78	1992	concrete	office
	Federation Towers - Vostok Tower	Moscow	373	1,224	95	2015	concrete	residential / hotel / office
	Oberoi Oasis Tower B	Mumbai	372	1,220	82	2016	concrete	residential
	The Address The BLVD	Dubai	370	1,214	72	2016	concrete	residential / hotel
	Golden Eagle Tiandi Tower A	Nanjing	368	1,208	76	-		hotel / office
	Chang Fu Jin Mao Tower	Shenzhen	368	1,207	68	2016	composite	office
23	Bank of China Tower	Hong Kong	367	1,205	72	1990	composite	office
24	Bank of America Tower	New York City	366	1,200	55	2009	composite	office
	Dalian International Trade Center	Dalian	365	1,199	86	2015	composite	residential / office
	VietinBank Business Center Office Tower	Hanoi	363	1,191	68	2017	composite	office
25	Almas Tower	Dubai	360	1,181	68	2008	concrete	office
25	The Pinnacle	Guangzhou	360	1,181	60	2012	concrete	office
27	JW Marriott Marquis Hotel Dubai Tower 1	Dubai	355	1,166	82	2012	concrete	hotel
27	JW Marriott Marquis Hotel Dubai Tower 2	Dubai	355	1,166	82	2013	concrete	hotel
29	Emirates Tower One	Dubai	355	1,163	54	2000	composite	office
	OKO - South Tower	Moscow	352	1,155	85	2015	concrete	residential / hotel
	Forum 66 Tower 2	Shenyang	351	1,150	68	2015	composite	office
	Hanking Center	Shenzhen	350	1,148	65	2018		office
	Spring City 66	Kunming	350	1,148	-	2018		office
	J97	Changsha	349	1,146	97	2014	steel	residential / office
30	Tuntex Sky Tower	Kaohsiung	348	1,140	85	1997	composite	hotel / office
31	Aon Center	Chicago	346	1,136	83	1973	steel	office
32	The Center	Hong Kong	346	1,135	73	1998	steel	office
33	John Hancock Center	Chicago	344	1,128	100	1969	steel	residential / office
	Four Seasons Place	Kuala Lumpur	343	1,124	65	2017		residential / hotel
	ADNOC Headquarters	Abu Dhabi	342	1,122	76	2014	concrete	office
	Ahmed Abdul Rahim Al Attar Tower	Dubai	342	1,122	76	2014	steel / concrete	residential
	Xiamen International Centre	Xiamen	340	1,115	61	2016	composite	office
	LCT Residential Tower A	Busan	339	1,113	85	2018		residential
34	The Wharf Times Square 1	Wuxi	339	1,112	68	2014	composite	hotel / residential
	Chongqing World Financial Center	Chongqing	339	1,112	72	2014	composite	office
35	Mercury City Tower	Moscow	339	1,112	75	2013	concrete	residential / office
	Tianjin Modern City	Tianjin	338	1,109	65	2016	composite	residential / hotel

* estimated height

Rank	Building Name	City	Height (meters)	Height (feet)	Stories	Year	Material	Function
	Orchid Crown Tower A	Mumbai	337	1,106	75	2016	concrete	residential
	Orchid Crown Tower B	Mumbai	337	1,106	75	2016	concrete	residential
36	Tianjin World Financial Center	Tianjin	337	1,105	75	2011	composite	office
37	The Torch	Dubai	337	1,105	79	2011	concrete	residential
38	Keangnam Hanoi Landmark Tower	Hanoi	336	1,102	72	2012	concrete	hotel / residential / office
	Wilshire Grand Tower	Los Angeles	335	1,100	73	2017	composite	hotel / office
	DAMAC Heights	Dubai	335	1,099	86	2016	steel / concrete	residential
39	Shimao International Plaza	Shanghai	333	1,094	60	2006	concrete	hotel / office
	LCT Residential Tower B	Busan	333	1,093	85	2018		residential
	Mandarin Oriental Hotel	Chengdu	333	1,093	88	2017		residential / hotel
40	Rose Rayhaan by Rotana	Dubai	333	1,093	71	2007	composite	hotel
	Jinan Center Financial City	Jinan	333	1,093	-	-		-
	China Chuneng Tower	Shenzhen	333	1,093	-	2017	composite	-
41	Minsheng Bank Building	Wuhan	331	1,086	68	2008	steel	office
	Ryugyong Hotel	Pyongyang	330	1,083	105	-	concrete	hotel / office
	Gate of Kuwait Tower	Kuwait City	330	1,083	84	2016	concrete	hotel / office
42	China World Tower	Beijing	330	1,083	74	2010	composite	hotel / office
	Thamrin Nine Tower 1	Jakarta	330	1,083	71	-		office
	Zhuhai St Regis Hotel & Office Tower	Zhuhai	330	1,083	67	2016	composite	hotel / office
	The Skyscraper	Dubai	330	1,083	66	-		office
	Yuexiu Fortune Center Tower 1	Wuhan	330	1,083	66	2016	composite	office
	Suning Plaza Tower 1	Zhenjiang	330	1,082	77	2016	composite	-
	Hon Kwok City Center	Shenzhen	329	1,081	80	2015	composite	residential / office
43	Longxi International Hotel	Jiangyin	328	1,076	72	2011	composite	residential / hotel
43	Al Yaqoub Tower	Dubai	328	1,076	69	2013	concrete	hotel
	Nanjing World Trade Center Tower 1	Nanjing	328	1,076	69	2016	composite	hotel / office
	Golden Eagle Tiandi Tower B	Nanjing	328	1,076	68	-		office
	Wuxi Suning Plaza 1	Wuxi	328	1,076	68	2014	composite	hotel / office
	Concord International Centre	Chongqing	328	1,076	62	2016	composite	hotel / office
	Baoneng Shenyang Global Financial Centre Tower 2	Shenyang	328	1,076	-	2018		hotel / office
	Greenland Center Tower 1	Qingdao	327	1,074	74	2016	composite	hotel / office
	Huaqiang Golden Corridor City Plaza Main Tower	Shenyang	327	1,073	66	2018		hotel / offiec
	Salesforce Tower	San Francisco	326	1,070	61	2017		office
45	The Index	Dubai	326	1,070	80	2010	concrete	residential / office
	Cemindo Tower	Jakarta	325*	1,066*	63	2015	concrete	hotel / office
46	The Landmark	Abu Dhabi	324	1,063	72	2013	concrete	residential / office
46	Deji Plaza	Nanjing	324	1,063	62	2013	composite	hotel / office
	Yantai Shimao No. 1 The Harbour	Yantai	323	1,060	59	2015	composite	residential / hotel / office
48	Q1 Tower	Gold Coast	323	1,058	78	2005	concrete	residential
	Lamar Tower 1	Jeddah	322	1,056	70	2016	concrete	residential / office
49	Wenzhou Trade Center	Wenzhou	322	1,056	68	2011	concrete	hotel / office
	Guangxi Finance Plaza	Nanning	321	1,053	68	2016	composite	hotel / office
50	Burj Al Arab	Dubai	321	1,053	56	1999	composite	hotel
51	Nina Tower	Hong Kong	320	1,051	80	2006	concrete	hotel / office
	Sinar Mas Center 1	Shanghai	320	1,048	66	2015	composite	office
52	Chrysler Building	New York City	319	1,046	77	1930	steel	office
	Global City Square	Guangzhou	319	1,046	67	2015	composite	office
53	New York Times Tower	New York City	319	1,046	52	2007	steel	office
	Foshan Suning Plaza Tower 1	Foshan	318	1,043	90	-		hotel / office
	Runhua Global Center 1	Changzhou	318	1,043	72	2015	composite	office

Rank	Building Name	City	Height (meters)	Height (feet)	Stories	Year	Material	Function
	Jiuzhou International Tower	Nanning	318	1,043	71	2016	composite	-
	Riverside Century Plaza Main Tower	Wuhu	318	1,043	66	2015	composite	hotel / office
54	**HHHR Tower**	**Dubai**	**318**	**1,042**	**72**	**2010**	**concrete**	**residential**
	Yurun International Tower	Huaiyin	317	1,040	75	2017	composite	office
	Chongqing IFS T1	Chongqing	317	1,038	64	2016	composite	hotel / office
	Namaste Tower	Mumbai	316	1,037	63	2017	concrete	hotel / office
	Changsha IFS Tower T2	Changsha	315	1,033	-	2017	composite	office
	Youth Olympics Center Tower 1	Nanjing	315	1,032	68	2015	composite	-
	MahaNakhon	Bangkok	313	1,028	77	2016	concrete	residential / hotel
	The Stratford Residences	Makati	312	1,024	74	2015	concrete	residential
55	**Bank of America Plaza**	**Atlanta**	**312**	**1,023**	**55**	**2014**	**composite**	**office**
	Moi Center Tower A	Shenyang	311	1,020	75	2014	composite	hotel / office
56	**US Bank Tower**	**Los Angeles**	**310**	**1,018**	**73**	**1990**	**steel**	**office**
57	**Ocean Heights**	**Dubai**	**310**	**1,017**	**83**	**2010**	**concrete**	**residential**
57	**Menara Telekom**	**Kuala Lumpur**	**310**	**1,017**	**55**	**2001**	**concrete**	**office**
	Bodi Center Tower 1	Hangzhou	310	1,017	55	2017		office
	Fortune Center	Guangzhou	309	1,015	73	2015	composite	office
59	**Pearl River Tower**	**Guangzhou**	**309**	**1,015**	**71**	**2012**	**composite**	**office**
	Poly Pazhou C2	Guangzhou	309	1,015	61	2017	composite	office
60	**Emirates Tower Two**	**Dubai**	**309**	**1,014**	**56**	**2000**	**concrete**	**hotel**
	Eurasia	Moscow	309	1,013	72	2014	composite	hotel / office
	Guangfa Securities Headquarters	Guangzhou	308	1,010	62	2016		office
61	**Burj Rafal**	**Riyadh**	**308**	**1,010**	**68**	**2014**	**concrete**	**residential / hotel**
	Wanda Plaza 1	Kunming	307	1,008	67	2016	composite	office
	Wanda Plaza 2	Kunming	307	1,008	67	2016	composite	office
	Lokhandwala Minerva	Mumbai	307	1,007	83	2015	concrete	residential
62	**Franklin Center - North Tower**	**Chicago**	**307**	**1,007**	**60**	**1989**	**composite**	**office**
63	**Cayan Tower**	**Dubai**	**306**	**1,005**	**73**	**2013**	**concrete**	**residential**
	One57	New York City	306	1,005	79	2014	steel / concrete	residential / hotel
64	**East Pacific Center Tower A**	**Shenzhen**	**306**	**1,004**	**85**	**2013**	**concrete**	**residential**
64	**The Shard**	**London**	**306**	**1,004**	**73**	**2013**	**composite**	**residential / hotel / office**
66	**JPMorgan Chase Tower**	**Houston**	**305**	**1,002**	**75**	**1982**	**composite**	**office**
67	**Etihad Towers T2**	**Abu Dhabi**	**305**	**1,002**	**80**	**2011**	**concrete**	**residential**
68	**Northeast Asia Trade Tower**	**Incheon**	**305**	**1,001**	**68**	**2011**	**composite**	**residential / hotel / office**
69	**Baiyoke Tower II**	**Bangkok**	**304**	**997**	**85**	**1997**	**concrete**	**hotel**
70	**Wuxi Maoye City - Marriott Hotel**	**Wuxi**	**304**	**997**	**68**	**2014**	**composite**	**hotel**
71	**Two Prudential Plaza**	**Chicago**	**303**	**995**	**64**	**1990**	**concrete**	**office**
	Diwang International Fortune Center	Liuzhou	303	994	75	2015	composite	residential / hotel / office
	KAFD World Trade Center	Riyadh	303	994	67	2015	concrete	office
	Jiangxi Nanchang Greenland Central Plaza 1	Nanchang	303	994	59	2015	composite	office
	Jiangxi Nanchang Greenland Central Plaza 2	Nanchang	303	994	59	2015	composite	office
72	**Leatop Plaza**	**Guangzhou**	**303**	**993**	**64**	**2012**	**composite**	**office**
73	**Wells Fargo Plaza**	**Houston**	**302**	**992**	**71**	**1983**	**steel**	**office**
74	**Kingdom Centre**	**Riyadh**	**302**	**992**	**41**	**2002**	**steel / concrete**	**residential / hotel / office**
75	**The Address**	**Dubai**	**302**	**991**	**63**	**2008**	**concrete**	**residential / hotel**
	Gate of the Orient	Suzhou	302	990	68	2014	composite	residential / hotel / office
76	**Capital City Moscow Tower**	**Moscow**	**302**	**990**	**76**	**2010**	**concrete**	**residential**
	Greenland Puli Center	Jinan	301	988	61	2015	composite	residential / office
	Greenland Center North Tower	Yinchuan	301	988	58	-	composite	hotel / office
	Heung Kong Tower	Shenzhen	301	987	70	2014	composite	hotel / office

* estimated height

Rank	Building Name	City	Height (meters)	Height (feet)	Stories	Year	Material	Function
	Brys Buzz	Greater Noida	300	984	82	2017	concrete	residential
77	Doosan Haeundae We've the Zenith Tower A	Busan	300	984	80	2011	concrete	residential
	Supernova Spira	Noida	300	984	80	2017	concrete	residential
	Al Habtoor City Tower 1	Dubai	300+*	984+*	74	2017	concrete	residential
	Al Habtoor City Tower 2	Dubai	300+*	984+*	74	2017	concrete	residential
	NBK Tower	Kuwait City	300	984	70	2017	concrete	office
	Huachuang International Plaza Tower 1	Changsha	300	984	66	2016	composite	hotel / office
	Riverfront Times Square	Shenzhen	300	984	64	2016	composite	hotel / office
77	Torre Costanera	Santiago	300	984	62	2014	concrete	office
77	Abeno Harukas	Osaka	300	984	62	2014	steel	hotel / office / retail
	Golden Eagle Tiandi Tower C	Nanjing	300	984	60	-		office
77	Arraya Tower	Kuwait City	300	984	60	2009	concrete	office
	Shenglong Global Center	Fuzhou	300	984	57	2017	composite	office
77	Aspire Tower	Doha	300	984	36	2007	composite	hotel / office
	Shum Yip Upperhills Tower 2	Shenzhen	300	984	-			office
	Jin Wan Plaza 1	Tianjin	300	984	66	2017		hotel / office
	Langham Hotel Tower	Dalian	300	983	74	2015	composite	residential / hotel
82	First Canadian Place	Toronto	298	978	72	1975	steel	office
82	One Island East	Hong Kong	298	978	68	2008	concrete	office
	Ilham Baru Tower	Kuala Lumpur	298	978	64	2015	concrete	residential / office
	Yujiapu Yinglan International Finance Center	Tianjin	298	978	60	2016	composite	office
84	4 World Trade Center	New York City	298	977	65	2014	composite	office
85	Eureka Tower	Melbourne	297	975	91	2006	concrete	residential
	Dacheng Financial Business Center Tower A	Kunming	297	974	-	2016	steel	hotel / office
86	Comcast Center	Philadelphia	297	974	57	2008	composite	office
87	Landmark Tower	Yokohama	296	972	73	1993	steel	hotel / office
88	R&F Yingkai Square	Guangzhou	296	972	66	2014	composite	residential / hotel / office
89	Emirates Crown	Dubai	296	971	63	2008	concrete	residential
	Xiamen Shimao Cross-Strait Plaza Tower B	Xiamen	295	969	67	2015	composite	office
90	Khalid Al Attar Tower 2	Dubai	294	965	66	2011	concrete	hotel
	Lamar Tower 2	Jeddah	293	961	62	2016	concrete	residential / office
91	311 South Wacker Drive	Chicago	293	961	65	1990	concrete	office
	Shang Xinguo International Plaza	Chongqing	293	961	65	-		hotel / office
92	Sky Tower	Abu Dhabi	292	959	74	2010	concrete	residential / office
93	Haeundae I Park Marina Tower 2	Busan	292	958	72	2011	composite	residential
94	SEG Plaza	Shenzhen	292	957	71	2000	concrete	office
	Indiabulls Sky Suites	Mumbai	291	955	75	2015	concrete	residential
95	70 Pine Street	New York City	290	952	67	2014	steel	residential / hotel
	Hunter Douglas International Plaza	Guiyang	290	951	69	2014	composite	hotel / office
	Tanjong Pagar Centre	Singapore	290	951	68	2016	composite	residential / hotel / office
	Powerlong Center Tower 1	Tianjin	290	951	59	2015	composite	office
	220 Central Park South	New York City	290	950	66	2017		residential
	Zhengzhou Eastern Center North Tower	Zhengzhou	289	948	78	2016	composite	office
	Zhengzhou Eastern Center South Tower	Zhengzhou	289	948	78	2016	composite	office
96	Dongguan TBA Tower	Dongguan	289	948	68	2013	composite	hotel / office
	Busan International Finance Center Landmark Tower	Busan	289	948	63	2014	concrete	office
97	Key Tower	Cleveland	289	947	57	1991	composite	office
98	Shaoxing Shimao Crown Plaza	Shaoxing	288	946	60	2012	composite	hotel / office
99	Plaza 66	Shanghai	288	945	66	2001	concrete	office
100	One Liberty Place	Philadelphia	288	945	61	1987	steel	office

Index of Buildings

4 World Trade Center, *New York City;* 36
6 Bevis Marks, *London;* 148
8 Chifley, *Sydney;* 64
10 Brock Street, *London;* 150
22 Rothschild Tower, *Tel Aviv;* 186
41X, *Melbourne;* 96
171 Collins Street, *Melbourne;* 98
500 Lake Shore Drive, *Chicago;* 50
1812 North Moore, *Arlington;* 50

Abeno Harukas, *Osaka;* 68
Academic 3, *Hong Kong;* 100
Albert Tower, *Melbourne;* 102
Anhui New Broadcasting & TV Center, *Hefei;* 104
Ardmore Residence, *Singapore;* 72
ASE Centre Chongqing R2, *Chongqing;* 130
Asia Square, *Singapore;* 130
AvB Tower, *The Hague;* 152

Baku Flame Towers, *Baku;* 106
Bloomberg Tower, *New York City;* 212
BSR 3, *Tel Aviv;* 178

CalypSO, *Rotterdam;* 154
Capital, The, *Mumbai;* 134
Cayan Tower, *Dubai;* 172
Champion Tower, *Tel Aviv;* 180
Changzhou Modern Media Center, *Changzhou;* 108
China Merchants Tower, *Shenzhen;* 110
China Resources Building, *Hong Kong;* 131
Concord Cityplace Parade, *Toronto;* 51
Conrad Hotel, *Dubai;* 186
Courtyard & Residence Inn, *New York City;* 51
Couture, *Toronto;* 52

Darling Quarter, *Sydney;* 236
DBS Bank Tower, *Jakarta;* 131
DC Tower, *Vienna;* 144
De Rotterdam, *Rotterdam;* 138

E' Tower, *Eindhoven;* 168
Edith Green-Wendell Wyatt Federal Building, *Portland;* 22
Exzenterhaus Bochum, *Bochum;* 156

Fake Hills, *Beihai;* 112
FKI Tower, *Seoul;* 76
Fletcher Hotel Amsterdam, *Amsterdam;* 158
Fortune Plaza Phase III, *Beijing;* 132

Gloucester, The, *Hong Kong;* 135
Godfrey, The, *Chicago;* 46
Gramercy Residences, *Makati;* 200
Grand Office, *Vilnius;* 168
Guangzhou Circle, *Guangzhou;* 114

Habitat, *Melbourne;* 116
Highlight Towers, *Munich;* 211

IDEO Morph 38, *Bangkok;* 80
Infinity, *Brisbane;* 132
Interlace, The, *Singapore;* 190
International Commerce Center, *Hong Kong;* 230

Jin Mao Tower, *Shanghai;* 234
Jinao Tower, *Nanjing;* 118
Jinling Hotel Asia Pacific Tower, *Nanjing;* 133
Jockey Club Innovation Tower, The, *Hong Kong;* 88
John and Frances Angelos Law Center, The, *Baltimore;* 53

K2 at K Station, *Chicago;* 52
Kent Vale, *Singapore;* 120

L'Avenue, *Shanghai;* 122
Landmark, The, *Abu Dhabi;* 184
Leadenhall Building, The, *London;* 220, 222

Magma Towers, *Monterrey;* 38
Martin, The, *Seattle;* 226
Maslak Spine Tower, *Istanbul;* 160
MuseumHouse, *Toronto;* 40

NBF Osaki Building, *Tokyo;* 216
NEMA, *San Francisco;* 53
NEO Bankside, *London;* 196

OLIV, *Hong Kong;* 124
One AIA Financial Center, *Foshan;* 133
One Angel Square, *Manchester;* 169
One Central Park, *Sydney;* 58

Pakubuwono Signature, The, *Jakarta;* 135
Peninsula Tower, *Mexico City;* 42
Peter Gilgan Centre, The, *Toronto;* 54
Point, The, *Guayaquil;* 28
Portside, *Cape Town;* 182
Post Tower, *Bonn;* 204

RMIT Swanston Academic Building, *Melbourne;* 134
Rosewood Abu Dhabi, *Abu Dhabi;* 187

Shanghai Arch, *Shanghai;* 126
Sheraton Huzhou Hot Spring Resort, *Huzhou;* 84
Solaria, *Milan;* 162
Solea, *Milan;* 169

Taipei 101, *Taipei;* 210
Territoria El Bosque, *Santiago;* 44
Time Warner Center, *New York City;* 212
Torre Agbar, *Barcelona;* 210
Torre Costanera, *Santiago;* 48
Torres del Yacht, *Buenos Aires;* 54
Tour Carpe Diem, *Paris;* 166
Tower Palace Three, *Seoul;* 213
Tower, The, One St George Wharf, *London;* 164

United Nations Secretariat Building, *New York City;* 32
Uptown Munchen, *Munich;* 211

Wangjing SOHO, *Beijing;* 92
World Trade Center Doha, *Doha;* 187

Xiamen Financial Centre, *Xiamen;* 128

YooPanama Inspired by Starck, *Panama City;* 55

ZenCity, *Buenos Aires;* 55

Index of Companies

3DReid; 169
4 Estaciones; 42
206 Bloor Street West Development Corporation; 40

AB Concept Limited; 135
Absolute Project Management; 182
ABT Delft; 139
Accura Systems; 44
ACLA; 100, 132
Acoustic Design Studio; 23
Acoustic Logic; 102, 132
ACSET; 135
Aciron Acoustics Consultants Pte Ltd; 130, 191
Ada Bronfman Engineers & Consultants Ltd; 180
Adamson Associates; 36, 212
Adana Dinamik Mühendislik; 106
ADG; 132
AD+RG Architecture Design and Research Group; 89
Adriana Hoyos; 29
Adrian Norman Limited; 135
Adrian Smith + Gordon Gill Architecture; 52, 77
Advance Mechanical Systems, Inc.; 52
AEC Limited; 135
AECOM; 96, 130, 134, 182, 187
AECOM / Davis Langdon; 53, 59, 96, 120, 150, 164, 184
Aedas; 133
Aercoustics Engineering; 54
AEV Topografia; 44
Agama Energy; 182
AGC Design; 89
AHF.sa Ingenieros Estructurales; 55
AIA Group Limited; 133
AIK; 59
Airmas Asri; 135
Alan G. Davenport Wind Engineering Group – BLWTL; 51, 173, 184, 234
Aldaag Engineers Consultants, Ltd.; 178
Alemparte, Barreda y Asociados; 48
Al Habtoor Engineering; 184
Allied Real Estate, Ltd.; 180
Alonso Larrin Evaristo; 44
Alpha Consulting, Ltd.; 131
Alsop Architects; 154
ALT Cladding; 48, 126, 130, 210
AMproject; 114
Anahuac Organización Constructora; 38
Ananda Development PCL; 81
Angeli; 54
Anhui Broadcasting & TV Station; 104
Anna van Buerenplein BV; 152
Anteus Constructora; 42
Aon Fire Protection Engineering Corporation; 53
A. Papish & Co. Consulting Engineering, Ltd.; 178
Apollo Real Estate Advisors, L.P.; 212
APP Corporation; 98
Applied Acoustics; 148
Applied Landscape Design; 184
Arabian Construction; 187
Arabtec; 173, 186, 187
Architects 61; 73, 130
Architectural Design & Research Institute of Guangdong Province, The; 110
Architecture & Access; 134
Architektūros Linija; 168
Ariatta Ingegneria; 162, 169
Arkonin; 131
Arnold Associates; 126
Arquitectonica; 162
Arte Baumanagement GmbH; 156

Arts Towers Development S.A.; 55
Arup; 59, 65, 73, 89, 93, 110, 128, 131, 132, 139, 150, 162, 169, 200, 220, 222, 231, 236
ASE Group; 130
Asia Square Tower Pte., Ltd.; 130
Aspect Oculus; 59
Aspect Studios; 65
Atakar Engineering; 160
Ateliers Jean Nouvel; 59, 210
Atelier Ten; 166
Atkins; 186
Aurecon; 96, 133, 182, 236
Auscoast Fire Services; 132
Australian Institute of Architects; 96
Aviva France; 166
Aviv & Co., Ltd.; 186
AXA; 148
Axima Sistemas e Instalaciones S.A.; 210
AXIS Ingenieursleistungen ZT GmbH; 145
Ayers Saint Gross; 53
Azinko Development MMC; 106

b720 arquitectos; 210
BA Group; 54
Balkar Mühendislik; 106
BAM Construction; 169
Bar Akiva Engineers; 180
Barton Wilmore; 164
Bates Smart; 98
BECA Group; 131, 186
Behnisch Architekten; 53
Beihai Xinpinguangyang Real Estate Development Co., Ltd.; 112
Beijing Institute of Architectural Design; 128
Beijing Special Engineering Design and Research Institute; 104, 220, 224, 226
Bekaert; 226
Belt Collins & Associates; 131, 231
Benatar Consulting; 182
Benthem Crouwel Architects; 158
Benway Limited; 124
Besix; 166
Bettis Tarazi Arquitectos S.A.; 55
BFBA; 182
BG&E Façades; 98
BIAD Lighting Design Studio; 128
Bill Rooney Studio, Inc.; 51
BlackRock; 148
BlackRock Property Singapore Ptd., Ltd.; 130
BLVD; 110
BMT Fluid Mechanics, Ltd.; 164, 187
Bodas, Miani, Anger Arquitectos & Asociados; 55
Boele & van Eesteren; 154
Bollinger + Grohmann; 145
Bonacci Group; 116, 132, 134
Bonbori Lighting Architect & Associates, Inc.; 69
Bo Steiber Lighting Design; 134, 186
BPK Brandschutzplanung Klingsch; 211
Brandi Consult GmbH; 205
Brandston Partnership, Inc.; 110, 118, 130
Brennan Beer Gorman Monk; 187
British Land; 150
Broadway Malyan; 164
Brookfield Multiplex; 98, 134, 164
B.S.R. Group; 178
Building System and Diagnostics; 130
Buro Happold; 169, 184
Buro Ole Scheeren; 191
BVDA; 51
bwp Burggraf + Reiminger Beratende Ingenieure GmbH; 211

Cabinet Daniel Legrand; 166
Cabinet Penicaud (now part of SNC Lavalin); 166
CABR; 93
Cadman; 226
Cagley & Associates; 53
Callison; 226
Cambridge University; 220
Campbell Shillinglaw Lau, Ltd.; 133
CapitaLand Singapore Limited; 191
Caputo Partnership; 169
Caputo S.A. & Xapor S.A.; 55
Cardno; 116
Carillion PLC; 197
Carter Family, The; 102
Cary Kopczynski & Company; 226
Cauberg-Huygen Consulting Engineers; 154
Cayan Investment & Development; 173
CB Engineers; 53
Cbus Property; 98
CCW Associates Pte., Ltd.; 131
CDML Consulting, Ltd.; 54
Cencosud S.A.; 48
Century Properties Group, Inc.; 200
Cerami Associates; 173
CERI, Ltd.; 132
Certis; 132
CH2M Hill; 150
Chang-Jo Architects; 77
Changzhou Broadcasting Station; 108
Changzhou Radio and TV Realty Company, Ltd.; 108
Charles M. Salter Associates; 23
Charter Hall; 98
Charter Sills & Associates; 52
Chatfield Electric, Inc.; 52
Chhada Siembieda Leung, Ltd.; 133
China Academy of Building Research; 104
China Construction Design International; 93
China Construction Eighth Engineering Division; 122
China Construction Third Engineering Bureau Co., Ltd.; 108, 133
China Jin Mao Group Co. Ltd; 234
China Majesty Steel Structural Design Co., Ltd.; 85
China Merchants Real Estate Shenzhen Co., Ltd.; 110
China Resources Property, Ltd.; 131
China State Construction Engineering Corporation; 93, 128, 133
Christian Wiese Architects; 29
Churba / Bernardini; 55
Ciputra Group; 131
CISDI; 130
Citicapital; 42
Cityplan Services; 132
City University of Hong Kong; 100
CL3 Architects, Ltd.; 131
Claassen Auret International; 182
Cladtech Associates; 164
Clark Construction; 50
Cline Bettridge Bernstein Lighting Design, Inc.; 187
Clive Newsome; 182
CMB; 162, 169
Code Consultants Professional Engineers, PC; 36
Coheco; 29
Coima Image; 162, 169
Cole Jarman; 164
Columbus Centre LLC; 212
Conco; 226
Concord Adex; 51
Consis Engineering; 200

Consolidated Contractors International Company; 184
Construction Cost Systems; 77
Constructora Sigro S.A.; 44
Consuambiente; 29
Continental Engineering Consultants, Inc.; 210
Contour Consultants Australia; 116
Co-operative Group, The; 169
CORE; 148
Corsmit Raadgevende Ingenieurs; 139
Cosentini; 50, 212
Coyado; 29
CPP; 54, 55
CR Construction Co., Limited; 131
Crédit Agricole Assurances; 166
CS Caulking Co., Inc.; 118
Currie & Brown; 173
Cutler Anderson Architects; 23
C.Y. Lee & Partners Architects/Planners; 210

Daccord; 46
Dagesh Engineering Traffic & Road Design, Ltd.; 178, 180, 186
Dai Nippon Construction; 69
Dakno S.A.; 55
David Engineers, Ltd.; 178, 180
Davis, Carter, Scott, Ltd.; 50
Davy Sukamta & Partners; 135
dbHMS; 50
DBI Design Pty., Ltd.; 132
DCWC; 134
Deerns; 162, 169
De Leeuw group; 182
Dennis Lau & Ng Chun Man Architects & Engineers; 130, 135
Denton Corker Marshall; 130
De Rotterdam CV; 139
Desarrollo Aluminero Lea; 42
Deutsche Post AG; 205
Deutsche Post Bauen; 205
Device Logic; 59
Dewberry; 50
De Wilgen Vasrgoed; 154
DGMR Raadgevende Ingenieurs; 139, 152
dhk architects; 182
DIA Holdings; 106
Diamond Schmitt Architects; 54
di Domenico + Partners, LLP; 33
Diversified Engineering; 53
Dominique Perrault Architecture; 145
DongYang Structural Engineers Co., Ltd.; 77
Doron Shachar Engineers; 180
Dot Line Plane; 81
DP9; 197
DPPS Projects; 96
Dragados; 210
Drees & Sommer Advanced Building Technologies; 156
Dr. Pfeiler GmbH; 145
DS-Plan GmbH; 211
dS+V; 154
dtah; 54
Dubai Contracting Company; 186
Ducibella Venter & Santore; 36
Durst Organization, The; 241
DVP; 139

Earth Asia; 130
Earthscape Concepts; 130
East China Architectural Design & Research Institute; 126, 234
EBV Fire & Security; 44
EC Harris; 187, 197
Ecoland; 93
EDAW; 110
Edgett Willams Consulting Group, Inc.; 118, 234
EDINTAR construcciones S.A.; 54
EDSA; 85

Edward & Zuck; 51
Eipeldauer & Partner GmbH; 145
Elemac Company limited; 81
ELIN GmbH & co KG; 145
EllisDon Corporation; 54
El-Rom Consulting Engineer, Ltd.; 178
Elsan Consultancy; 160
E. Lynn - S. Cohen Plumbing Consultant, Ltd.; 186
EMEPA; 54
EMSI; 93, 132
ENCO Energie-Consulting GmbH; 211
Energy 3Arq; 44
Engineering PLUS, LLC; 50
Eng. S. Lustig - Consulting Engineers, Ltd.; 186
Enrique Montoya Ingenieria, Ltda.; 44
Environmental Systems Design, Inc.; 77
EPADESA; 166
EPPAG; 184
Erdemli Engineering; 160
Eriksson Engineering Associates, Ltd.; 52
Eris Property Group; 182
Ernesto NA; 29
Estudio Grinberg GF S.A.; 54
Etik Consultancy; 160
Evergreen Engineering; 210
Exxel Pacific; 226
Exzenterhaus Bochum GmbH & Co. KG; 156

Fainstein AHF S.A.; 54
FBEYE International; 130
Feder Architects; 180
Federation of Korean Industries; 77
Feizhou Group; 85
Fernandez Prieto & Asoc. Ingenieros y Arquitectos S.A.; 54, 55
Fernandez Prieto Desarrollos Inmobiliarios S.A.; 54, 55
Ferris+Associates, Inc.; 51
Fideicomiso Puerto Madero 7; 55
Fifield Companies; 52
First Rand Bank; 182
Fisher Marantz Stone; 173
Fleischmann Ingenieria de Proyectos, Ltda.; 44
Fletcher Hotel Group; 158
Fletcher Priest Architects; 148
Flix Design; 81
Fortune Consultants, Ltd.; 77, 110, 126, 186
Foshan Main Forum Real Estate Development Company Limited; 133
Frances Krahe & Associates, Inc.; 106
Francis-Jones Morehen Thorp; 236
Frasers Property Australia; 59
Furigo; 55

Gallagher Jeffs; 102
Galotti Spa; 162, 169
Gardiner & Theobald Inc; 33, 169
GC Bankside LLP; 197
GCNY Group; 51
General Services Administration; 23
Genius Loci; 210
Gepro S.A.; 210
Gerald D Hines Interests; 211
Gerhard Spangenberg Architekt; 156
Gettys Group Hospitality Design; 46
Gill; 106
Gillespies; 197
GLR Arquitectos; 38
Gmeiner Haferl Zivilingenieure ZT GmbH; 145
G.N.B.A. Consultores S.R.L.; 55
Golder & Associates; 51, 132
Gormaz y Zenteno, Ltda.; 44
Gradient Microclimate Engineering, Inc.; 52
Granite Broadway Development, LLC; 51
Gravity Green Ltd.; 128
Gravity Partnership; 128
Graziani + Croazza Architects; 52
Green College Court BV; 152

Grontmij; 164
Grupo Inmobiliario Monterrey; 38
GSG Consultants, Inc.; 52
Guang Dong Hong Da Xing Ye Group; 114
Guang Dong Hong He Construction, Ltd.; 114
Guangdong Provincial Academy of Building Research; 110
GuD GmbH; 156
GV Ingenierie; 166

Hadi Komara; 131
Halcrow MWT; 132
Haley & Aldrich; 187
Handel Architects; 44, 53, 187
Hanil MEC. Engineering Co., Ltd; 77
Hantaran Prima Mandiri; 135
Harari LA; 38
Harbour Century Limited; 124
Harbour Vantage; 231
Harr Holand Consultancy; 160
HBO+EMTB Group; 134
HB Teknik; 106
HCE Engineers; 132
HDR Architecture; 54
Heavenward; 44
Heinle Wischer + Partners; 205
Henderson Land Development Company Limited; 135
H. Engineer Co., Ltd; 81
Heng Lai Construction Company Limited; 135
HGC Engineers; 52
H.H. Angus & Associates, Ltd.; 54
HHP Nord; 156
Hickory Group; 96
Hill International; 106
Hilson Moran; 184
Hines France; 166
Hines Italia; 162, 169
Hiromura Design Office; 69
HKI China Land, Ltd.; 132
HLW International LLP; 33
Hoare Lea; 197
HOCHTIEF Construction AG Niederlassung Hamburg; 205
Hoffmann-Janz Architekten; 145
HOK; 106
Hollingsworth Architects, LLC; 52
Hong Kong Polytechnic University, The; 89
Hospital for Sick Children, The; 54
Hotel Properties Limited; 191
Ho Wang & Partners Ltd.; 89
Howard S. Wright Construction; 23
HRVAC Consulting Engineering Co., Ltd.; 178
HS&A; 122
Hsin Chong Construction Company Limited; 100
Hua Nan Technology University; 110
Hyder Consulting; 100, 133
Hyundai Engineering & Construction; 77, 130

ICN Design International Pte. Ltd.; 191
Icon Construction Australia; 102
IECSA; 54
II By IV Design Associates, Inc.; 51
iki design group; 160
Ikonik; 93
Iluminación Sudamericana S.A.; 55
Imecanic; 29
Ing. Büro Landwehr GmbH; 156
ingenhoven architects; 211
Ingenieria Atlantico; 55
Ingeniería Carpenn; 55
Ingenieria y Controles Coyoacán; 42
Ingenieur Consult; 211
Ingenieurgesellschaft Heimann; 211
Ing. Eugenio Mendiguren; 55
Inhabit Group; 93
Inmo Mariuxi; 29
Insada Integrated Design Team; 135

Inspecta; 48
Instalaciones de Aire S.A. de C.V.; 42
Instalaciones Planificadas S.A. de C.V.; 42
Integrated Building Consultants; 130
Intelligent; 55
Interface; 23
International Project Management; 186
Isometrix; 184
Israel Berger & Associates; 51

J&A; 162, 169
Jablonsky Ast & Partners; 40
Jacobs Engineering; 162, 169
JAHN; 205, 211
James Law Cybertecture International; 134
Japsen Consultancy; 160
Jaros Baum & Bolles Consulting Engineers; 36
Jaya CM; 135
Jaya Konstruksi; 131
Jean-Claude HARDY; 59
Jean Jegou; 166
Jeff Bryar & Associates; 116
Jenkins & Huntington, Inc.; 50, 51, 52
Jenny Holzer; 65
Jeppe Aagaard Andersen; 59
Jerde Partnership Inc.; 200
Jiang Architects and Engineers; 112
Jiangsu Goldenland Real Estate Development
    (Group) Co., Ltd.; 118
Jiangsu Provincial Architectural D&R Institute,
    Ltd.; 133
Jinling Hotel Corporation, Ltd.; 133
John Portman & Associates; 126
JOOFarrill M Arquitectos S.A. de C.V.; 42
J. Roger Preston Group; 73, 120, 133, 135, 231
Jun Akoi & Associates; 122

Kai Shing Management Services Limited; 231
K.A.M.N. Structural Protection Consulting; 186
Kaplan Gehring McCarrol Architectural Lighting,
    Inc.; 131
Kardorff Ingenieure; 211
KCE Structural Engineers; 50
Kennovations; 59
Keppel Reit; 65
Khatib & Alami; 173
Kintetsu Corporation; 69
Kleinmann Engineering GmbH; 156
Kobi Gamzo; 180
Kohn Pedersen Fox Associates; 51, 162, 169, 231
Kolenik Eco Chic Design; 158
KPFF Consulting Engineers; 23
Krashin Electrical Engineering Consulting, Ltd.; 178
Kroll Inc; 33
KTP Consultants Pte., Ltd.; 120
Kumagai Gumi; 210

Labonia; 54
Laing O'Rourke; 220, 222
LAND; 162, 169
Land Art Design; 40
Langan Engineering & Environmental Services;
    51, 212
Langdon & Seah; 100, 131, 135, 191
Larden Muniak Consulting, Inc.; 51
Lasa; 166
Lea Consulting, Ltd.; 51
Leber / Rubes Inc.; 54
Lehr Associates; 210
Leigh and Orange Ltd.; 122
Lend Lease; 46, 50, 150, 212, 236
Lerch Bates; 77, 173, 210, 231
Leshem Sheffer Environmental Quality, Ltd.; 180
Leslie E. Robertson Associates; 36
LiftEye Ltd; 224
Lightdesign; 93
Lighting Planners Associates (S) Pte Ltd; 120, 191
Liora Niv Fromchenko; 180

Lippmann Partnership; 65
Living Design; 158
LM Liftmaterial GmbH; 224
Louis Karol Architects; 182
Lovell Chen; 98
L-Plan; 205
LR Group, Ltd.; 186
Lucciola – Fass Yakov; 54
Luis Bozzo Estructuras y Proyectos S.L.; 42
Lumina; 135
Lyons; 96, 134

M3 Consulting; 150
MAB; 139
MacKenzie Hoy Consulting Acoustic Engineers; 182
MAD Architects; 85, 112
Magnusson Klemencic Associates; 50, 53, 187
Mahimtura Consultants, Pvt., Ltd.; 134
Maki and Associates; 36
Mandiri Eka Abadi; 135
Mathematics and Mechanics Faculty of St
    Petersburg State University; 224
Matrix Consulting Services; 182
Maurice Brill Lighting Design; 150
Max Bögl Bauunternehmung GmbH & Co; 145
Maxcon Pty, Ltd.; 116
MBTW; 52
M. Caransa BV; 158
McClymont + RAK Engineers, Inc.; 52
McHugh Construction; 52
McKenzie Group; 116
MCLA Lighting Design; 50, 53
MCW Consultants, Ltd.; 51
M. Doron - I. Shahar & Co. Consulting Eng., Ltd.;
    186
Mect Co., Ltd.; 81
Meinhardt; 81, 131, 133, 134, 186, 187
MEL Consultants Pty., Ltd.; 102, 116, 132, 134
Merber Corporation; 40
Meriton Group; 132
Metal und Fassaden Consultancy; 160
Metrum Project Management; 182
M.G Acoustic Consultants, Ltd.; 180
Migdal group; 180
Mike Niven Interior Design; 51, 52
Mirvac Constructions; 65
Mirvac Developments; 65
MJR Management; 81
MKPL Architects Pte., Ltd.; 120
MMM Group; 51, 54
moBius consult; 158
Monarch Corporation; 52
Monarch Couture Developments, Ltd.; 52
Monday Properties; 50
Moshe Tzur Architects and Town Planners, Ltd.;
    186
M.SG.S.S.S. Arquitectos; 54
Mubadala Development Company; 187
Mueller Associates; 53
Murray & Roberts; 182
MuseumHouse Condominium Corporation; 40
Mutabilis; 166
MVA Transportation; 122, 135
MV Shore Associates, Ltd.; 52
MYS Architects; 180
MZ & Partners; 187

Nanjing Institute of Landscape Architecture Design
    & Planning, Ltd.; 133
National Center of Performing Center; 104
National University of Singapore; 120
Nelissen BV; 168
NET Project Management and Consultancy
    Services; 148
Newcomb & Boyd; 126
NewDesignArchitecture; 104
New Jinling Hotel Limited Company; 133
Nikken Sekkei; 217

Nobutaka Ashihara Architect, PC; 51
Northcroft Lim Consultants; 130
Nova Fire Protection; 46
Nusa Raya Cipta; 131

Obayashi Corporation; 69
Obiol, Moya i Associates; 210
Octatube; 158
Office for Metropolitan Architecture; 139, 191
OK Osadnik & Kamienski GmbH; 145
Okumura Corporation; 69
Old Mutual Properties; 182
Omer Construction & Engineering, Ltd.; 180
OM Ramirez & Associates S.A.; 55
Opening Solutions, Inc.; 173
Ortam – Malibu Group; 178
Oslund & Associates; 50
OVG Projectontwikkeling; 139
Oxford Capital Group, LLC; 46

PAE Consulting Engineers; 23
Page + Steele / IBI Group Architects; 40, 51
Pakubuwono Development, The; 135
Pamintori Cipta; 131
Pappageorge Haymes Partners; 52
Parsons Brinckerhoff; 132, 231
Patrick Blanc; 59
PCL; 51
Pelli Clarke Pelli Architects; 48, 69, 184, 212
Peridian Asia Pte., Ltd.; 186
Permasteelisa Group; 139, 231, 234
Perri Projects; 102
Peter Walker; 205
PLACE; 23
Planning & Management Consultants; 135
PLP; 134
Plumbing Corporation, Inc.; 55
Pomeroy Studio Pte Ltd; 200
Pontiac Land Group; 73
Private Property Management, Abu Dhabi; 186
Projacs International; 187
Proji-tech; 182
Pronobis; 29
Prüfstelle für Brandschutztechnik; 145
Pryme; 132
PT Ciputra Property, Tbk.; 131
P & T Group; 132, 133
PT Perentjana Djaja; 131
PTW Architects; 59

Qatar General Insurance & Reinsurance, Co.; 187

R.A. Heintges & Associates; 33
Ramboll; 148
Rami Ballas Engineering, Ltd.; 186
Raul A. Curuchet José Maria Del Villar Ingenieros
    Civiles; 55
R.Avis Surveying, Inc.; 52
Reddo; 102
Reed Jones Christoffersen Consulting Engineers; 52
REEF Associates Ltd; 164, 184
Related Companies, The; 212
Related Midwest; 50
René Lagos y Asociados; 48
Renzo Tonin & Associates; 65, 116
Residencial Peninsula Santa Fe; 42
Reynolds & Partners; 131
RFR; 187
Rider Levett Bucknall; 89, 98, 110
RK & K; 53
RMIT University; 134
Robert A.M. Stern Architects; 166
Robert Bird Group; 59, 164
Robert Brufau y Asociados; 210
Robert Pope Associates, Inc.; 50
Rogers Stirk Harbour + Partners; 65, 197, 222
Rolf Jensen & Associates; 33, 53, 77, 173
Ronald Lu & Partners; 100, 131

ROTHELOWMAN; 102, 116
RSEA Engineering; 210
RSP Architects, Planners & Engineers Pte Ltd.; 191
RTKL; 131
RTLD Lighting Design; 180
Ruscheway Consultancy; 160
Rush Wright Associates; 134
Ruz & Vukasovic; 44
RWDI; 48, 50, 51, 54, 77, 210, 212, 231, 247

Sahi Harpaz - Electrical Consulting & Eng., Ltd.; 186
Sako & Associates, Inc.; 173
Salfa Corporation; 48
Samartano & Company; 52
Samoo Architects & Engineers; 213
Samsung C&T Corporation; 210, 213
Sanfield Building Contractors Limited; 231
Santolaya Ingenieros Consultores; 44
Scheldebouw BV; 168
Schlaich Bergermann und Partner; 156
Schmidlin; 132
Schuler Shook; 46
S.D. Keppler & Associates, LLC; 36
Second Construction Co., Ltd. of China Construction Third Engineering Bureau, The; 104
Second Construction Engineering Co., Ltd., The; 110
Sekisui House Australia; 59
SERA Architects; 23
Servicios Portuarios; 55
Shanghai Citelum Lighting Design Co., Ltd.; 133
Shanghai Institute of Architectural Design & Research, Co., Ltd.; 108
Shanghai Jin Hong Qiao International Property Co., Ltd.; 126
Shanghai Jin Mao Contractor; 234
Shanghai Luxchina Property Development Co., Ltd.; 122
Shanghai No. 7 Construction Co., Ltd; 126
Shanghai Xian Dai Architecture Design (Group) Co., Ltd.; 85
Shen Milsom Wilke, Inc.; 33, 51, 53, 77, 173, 187, 210
Shen Zhen Jian Hong Da Construction, Ltd.; 114
Shen Zhen Rui Hua Construction, Ltd.; 114
Sheraton Huzhou Hot Spring Resort; 85
Shimizu Corporation; 73
Shiner + Associates; 46, 50, 52
Shma Company Limited; 81
Shui On Construction & Materials; 89
Siemens Building Technology; 210
SIGGMA; 114
Silverstein Properties; 36
SIP Project Managers; 182
Sitetectonix Pte., Ltd.; 120
Siu Yin Wai & Associates, Ltd.; 131
Skanska; 33, 148
Skidmore, Owings & Merrill; 110, 118, 173, 212, 213, 220, 234
SL+A International Asia; 210
Slattery Australia; 102
SLCE Architects; 212
S. Lustig Engineers & Consultants, Ltd.; 180
Smallwood, Reynolds, Stewart, Stewart; 186
S. Mashiah Consultants in Acoustics, Ltd.; 186
SMT Engineering; 160
SNC Lavalin; 166
Sociedad Inversora Dique IV S.A.; 54
Socsa; 38
SOHO China Co. Ltd; 93
Solomon Cordwell Buenz; 50
Solution Station; 182
Soma Group; 160
Somdoon Architects; 81
South China University of Technology; 114
Spectrum Design and Associates; 131

SPIE Batignolles; 166
Spoormaker and Partners; 182
Squire Mech Private Limited; 191
SRA Architectes; 166
Stadionkwartier BV; 168
Stahlform Baustahlbearbeitungs GmbH; 145
Stantec; 23, 52
STDM; 122
Stein Ltd; 224
Stephen Cheng Consulting Engineers Limited; 135
Stephen Stimson Associates; 53
St George South London; 164
STRABAG AG; 145, 211
Structural Affiliates International, Inc.; 46
Strukton Bouw; 158
Studio on Site; 69
Studio Raymond Chau Architecture Limited; 124
STX JV; 77
Suncorp Electrical; 132
Sun Hung Kai Properties; 231
Superior Air; 132
Surface Design Pty Ltd; 59
Susana Lihu Landscape Architecture; 178
Sustainable Design Consulting, LLC; 50
SWA Group; 77, 118, 173
Swinerton Builders; 53
Syasa Panama; 55
SYSKA Hennessy Group; 33

Taipei Financial Center Corporation; 210
Taiwan Kumagai; 210
Takenaka Corporation; 69
Talent Mechanical and Electrical Engineers, Ltd.; 131
TAP Consulting Engineers, Ltd.; 122
Tata; 131
Ta-You-Wei Construction; 210
Team 73 Hong Kong Ltd; 89
Techniplan Adviseurs; 139
Tenth and Market, LLC; 53
Teodoro González de León; 42
Terra Engineering, Ltd.; 52
Terrell Group; 166
Territoria; 44
Terry Hunziker; 73
TGM; 139
Thai Thai Engineering; 81
THEO TEXTURE; 124
Thomas Nicolas; 102
Thornton Tomasetti; 48, 77, 210, 212
Tielemans BV; 168
Tilaga; 160
Tiong Seng Constractors Pte., Ltd.; 120
Tishman Construction; 36
Townshend Landscape Architects; 148
Transsolar; 53, 59, 205
Trustful Engineering & Construction Co., Ltd.; 124
Turf Design; 59
TY Lin International; 191

Umow Lai; 98
United Nations; 33
University of Baltimore; 53
University of Michigan Civil and Environmental Engineering; 226
Uno Proyectos; 44
UN Studio; 73
Urbis; 98, 131, 133

V3 Companies; 46, 77
Valerio Dewalt Train Associates, Inc.; 46
Valstar Simonis; 139
Van Der Laan Bouma Architekten BV; 154
Van Deusen & Associates; 50, 173
Van Rossum Raadgevende Ingenieurs; 152, 158
Veisman Consulting Limited; 40
Vermessung Angst ZT GmbH; 145
Vicland Corporation Pty, Ltd.; 116

VIKA, Inc.; 50
Viltekta; 168
Viridian Energy & Environmental, LLC; 33, 187
Vulcan Real Estate; 226

Wacker Ingenieure; 145
WAC Projects; 182
Wadhwa Group, The; 134
Warren Thomas Plumbing Co.; 52
Waterman Building Services; 148
Waterman Group; 197
Watermark Associates; 133
Watkins Payne Partnership; 150
Watson Steel Structures Ltd; 220
Watt International; 48
Web Structures Singapore; 73
Weidlinger Associates; 33
Wells Fargo; 148
Werner Sobek Group; 145, 205
Westcon Co., Ltd; 81
Westwood Hong & Associates Ltd; 89
White Young Green; 164
Whiting-Turner Contracting Company, The; 53
Wichers & Dreef; 158
Wiel Arets Architects; 152, 168
Wiener Entwicklungsgesellschaft für den Donauraum AG; 145
Wilde & Woollard; 134
Wilkinson Eyre Architects; 150
Windtech Consultants; 131, 186
Winward Structures; 96, 98, 102
Wiratman & Associates; 131
W. J. Higgins & Associates, Inc.; 50
WMA Consulting Engineers, Ltd.; 46
Woh Hup Pte Ltd; 191
Wohr Parking Systems Pvt., Ltd.; 134
Wolff Landscape Architecture; 46
Wong & Cheng Consultants Engineers Limited; 124
Wong & Ouyang; 231
Wood and Grieve Engineers; 102
Wood Partners; 52
WSP Cantor Seinuk; 212
WSP Flack + Kurtz; 212
WSP Group; 51, 118, 122, 182, 187
WT Partnership; 132, 133, 148, 197, 231
Wuhan Construction Engineering Group Co., Ltd.; 118
Wuhan Lingyun; 104

Xiamen BIAD Architectural Design Co., Ltd.; 128
Xiamen Land Development Company; 128
Xiang Jiang Xing Li Estates Development, Ltd.; 132

Y.A. Yashar Architects, Ltd.; 178
YIT; 168
YIT Austria GmbH; 145
Yolles, a CH2M HILL Company; 51, 54
Yonsei University; 93
Yorkville Construction Corporation; 40

Zaha Hadid Architects; 89, 93
Zenitaka Corporation, The; 69
ZFG-Projekt; 145
Zhejiang Zhongnan Curtain Wall Co., Ltd.; 85
Zublin; 139
Zur Wolf Landscape Architects; 180

Image Credits

Front Cover: (from left to right) Edith Green-Wendell Wyatt Federal Building, © Nic Lehoux; One Central Park, © Murray Fredericks, courtesy of Frasers Property and Sekisui House; De Rotterdam, © OMA; photography by Michel van de Kar; Cayan Tower, © Tim Griffith

Pg 8: © Nic Lehoux

Pg 9: © HLW International LLP

Pg 10: © Somdoon Architects

Pg 11: One Central Park © John Gollings, courtesy of Frasers Property and Sekisui House; Abeno Harukas © Nakamichi Atsushi

Pg 12: De Rotterdam, © OMA; photography by Michel van de Kar; Sheraton Huzhou Hot Spring Resort © Sheraton Huzhou Hot Spring Resort

Pg 13: Ardmore Residence, © Pontiac Land Group; Cayan Tower, © Tim Griffith

Pg 14: © Iwan Baan

Pg 15: International Commerce Centre, © International Commerce Centre; Jin Mao Tower, (CC-by-SA) Shizao

Pg 16: Post Tower, © Marshall Gerometta; BioSkin, © Harunori Noda

Pg 22: © Nic Lehoux

Pg 23: © SERA Architects

Pg 24–27: Photos © Nic Lehoux; drawings © SERA Architects

Pg 28–31: All © Christian Wiese Architects

Pg 32–35: All © HLW International LLP

Pg 36–37: All © Silverstein Properties

Pg 38–39: All GLR © Arquitectos

Pg 40–41: All © Page+Steele/IBI Group Architects

Pg 42–43: All © Teodoro González de León

Pg 44: © Vincent Pouzet

Pg 45: Top image © Vincent Pouzet; bottom image © Natalia Vial; drawings © Handel Architects

Pg 46–47: Photos © Steve Hall, Hedrich Blessing; drawing © Valerio Dewalt Train Associates, Inc.

Pg 48–49: All © Pelli Clarke Pelli Architects

Pg 50: 500 Lake Shore Drive, © Hedrich Blessing; 1812 North Moore Street, © Eric Taylor

Pg 51: Concord Cityplace Parade, © Page+Steele/IBI Group Architects; Courtyard & Residence Inn Manhattan/Central Park, © ESTO

Pg 52: Couture, © Graziani + Croazza Architects; K2 at K Station, © Pappageorge Haymes Partners

Pg 53: NEMA, © NEMA; The John and Frances Angelos Law Center, © David Matthiessen

Pg 54: The Peter Gilgan Centre for Research and Learning, © Tom Arban; Torres del Yacht, © Fernandez Prieto Desarrollos Inmobiliarios S.A.

Pg 55: YooPanama Inspired by Starck, © Claudia Uribe; ZenCity © Fernandez Prieto Desarrollos Inmobiliarios S.A.

Pg 58: © Murray Fredericks, courtesy of Frasers Property and Sekisui House

Pg 59: © Simon Wood, courtesy of Frasers Property and Sekisui House

Pg 60: Left image © Simon Wood, courtesy of Frasers Property and Sekisui House; right image © John Gollings, courtesy of Frasers Property and Sekisui House

Pg 61–62: © Murray Fredericks, courtesy of Frasers Property and Sekisui House

Pg 63: © Ateliers Jean Nouvel

Pg 64–67: Photos © Brett Broadman; drawing © Mirvac Developments

Pg 68: © Hisao Suzuki

Pg 69–71: Photos © Nakamichi Atsushi; drawing © Takenaka Corporation

Pg 72–75: Photos © Iwan Baan; drawing © UN Studio

Pg 76–79: All © Adrian Smith + Gordon Gill Architecture

Pg 80–83: All © Somdoon Architects

Pg 84–87: All © MAD Architects

Pg 88: © Doublespace

Pg 89–91: Photos © Virgile Simon Bertrand; drawings © Zaha Hadid Architects

Pg 92: © Feng Chang

Pg 93: © Xia Zhi

Pg 94–95: Photo © Cao Baiqiang; drawings © Zaha Hadid Architects

Pg 96–97: Photos © John Gollings; drawing © Lyons

Pg 98–99: Photos © Peter Clarke; drawing © Charter Hall

Pg 100–101: All © Ronald Lu & Partners

Pg 102–103: Photos © Jaime Diaz-Berrio; drawings © ROTHELOWMAN

Pg 104–105: All © Anhui Broadcasting & TV Station

Pg 106–107: All © HOK

Pg 108–109: All © Shanghai Architectural Design Institute Co., Ltd.

Pg 110–111: Photos © Tim Griffith; drawing © Skidmore, Owings & Merrill LLP

Pg 112–113: All © MAD Architects

Pg 114: © AMproject

Pg 115: Top image © Guangdong Plastics Exchange; bottom image and drawings © AMproject

Pg 116–117: Photos © John Gollings; drawing © ROTHELOWMAN

Pg 118–119: Photos © Tim Griffith Photography; drawings © Skidmore, Owings & Merrill LLP

Pg 120–121: All © MKPL Architects Pte., Ltd

Pg 122–123: All © Leigh & Orange Ltd.

Pg 124–125: All © THEO TEXTURE

Pg 126–127: Photos © Luo Wen/VMA VISUAL; drawings © John Portman & Associates

Pg 128–129: All © Gravity Partnership, Ltd.

Pg 130: ASE Centre Chongqing R2, © Dennis Lau & Ng Chun Man Architects & Engineers, Ltd.; Asia Square, © John Gollings

Pg 131: China Resources Building, © Ronald Lu & Partners; DBS Bank Tower, © PT Ciputra Property, Tbk.

Pg 132: Fortune Plaza Phase III, © P & T Group; Infinity, © Meriton Group

Pg 133: Jinling Hotel Asia Pacific Tower, © P & T Group; One AIA Financial Center, © Aedas

Pg 134: RMIT Swanston Academic Building, © Lyons; The Capital, © James Law Cybertecture International Holdings, Ltd.

Pg 135: The Gloucester, © Dennis Lau & Ng Chun Man Architects & Engineers, Ltd.; The Pakubuwono Signature, © Airmas Asri

Pg 138: © OMA, by Richard John Seymour

Pg 139: Image courtesy of OMA, photography by Ossip van Duivenbode

Pg 140: © OMA, by Richard John Seymour

Pg 141: Top image courtesy of OMA, photography by Ossip van Duivenbode; bottom image © OMA, by Richard John Seymour

Pg 142: Image courtesy of OMA, photography by Philippe Ruault

Pg 143: All © OMA

Pg 144: © Michael Nagl/Dominique Perrault Architecture/Adagp

CTBUH Organizational Structure & Members

Saudi Binladin Group / ABC Division
Severud Associates Consulting Engineers, PC
Shanghai Construction (Group) General Co. Ltd.
Shree Ram Urban Infrastructure, Ltd.
Sinar Mas Group - APP China
Skanska
Solomon Cordwell Buenz
Studio Gang Architects
Syska Hennessy Group, Inc.
TAV Construction
Tongji Architectural Design (Group) Co., Ltd. (TJAD)
Walter P. Moore and Associates, Inc.
Werner Voss + Partner

Contributors
Aedas, Ltd.
Akzo Nobel
Allford Hall Monaghan Morris Ltd.
Alvine Engineering
Bates Smart
Benoy Limited
Bonacci Group
Boundary Layer Wind Tunnel Laboratory
Bouygues Construction
The British Land Company PLC
Canary Wharf Group, PLC
Canderel Management, Inc.
CBRE Group, Inc.
CCL
Continental Automated Buildings Association (CABA)
CTSR Properties Limited
DBI Design Pty Ltd
DCA Architects Pte Ltd
Deerns Consulting Engineers
DK Infrastructure Pvt. Ltd.
Dong Yang Structural Engineers Co., Ltd.
Far East Aluminium Works Co., Ltd.
GGLO, LLC
Goettsch Partners
Gradient Wind Engineering Inc.
Graziani + Corazza Architects Inc.
Hariri Pontarini Architects
The Harman Group
Hiranandani Group
Humphreys & Partners Architects, L.P.
Irwinconsult Pty., Ltd.
Israeli Association of Construction and Infrastructure Engineers (IACIE)
Jiang Architects & Engineers
JLL
KHP Konig und Heunisch Planungsgesellschaft
Langdon & Seah Singapore
LeMessurier
Lend Lease
Liberty Group Properties
Lusail Real Estate Development Company
M Moser Associates Ltd.
Mori Building Co., Ltd.
Nabih Youssef & Associates
National Fire Protection Association
National Institute of Standards and Technology (NIST)
National University of Singapore
Norman Disney & Young
OMA
Omrania & Associates
The Ornamental Metal Institute of New York
PACE Development Corporation PLC
Pei Partnership Architects
Perkins + Will
Pomeroy Studio Pte Ltd
PT Ciputra Property, Tbk
RAW Design Inc.
Ronald Lu & Partners
Royal HaskoningDHV
Sanni, Ojo & Partners
Shanghai Jiankun Information Technology Co., Ltd.
Silvercup Studios
SilverEdge Systems Software, Inc.
Silverstein Properties
SIP Project Managers Pty Ltd
The Steel Institute of New York
Stein Ltd.
SWA Group
SWA Group (Quarterly Mbr Mailing Address only)
Tekla Corporation
Terrell Group
TSNIIEP for Residential and Public Buildings
University of Illinois at Urbana-Champaign
Vetrocare SRL
Wilkinson Eyre Architects
Wirth Research Ltd
Woods Bagot

Participants
ACONSE CZ
ACSI (Ayling Consulting Services Inc)
Adamson Associates Architects
ADD Inc.
Aidea Philippines, Inc.
AIT Consulting
AKF Group, LLC
AKT II Limited
Al Jazeera Consultants
Alimak Hek AB
alinea consulting LLP
Alpha Glass Ltd.
ALT Cladding, Inc.
Altitude Façade Access Consulting
AMP Capital Investors
ARC Studio Architecture + Urbanism Pte. Ltd.
ArcelorMittal
Archetype Group
Architects 61 Pte., Ltd.
Architectural Design & Research Institute of Tsinghua University
Architectus

Arquitectonica
Atkins
Azorim Construction Ltd.
Azrieli Group Ltd.
Bakkala Consulting Engineers Limited
Baldridge & Associates Structural Engineering, Inc.
BAUM Architects
BDSP Partnership
Beca Group
Benchmark
BG&E Pty., Ltd.
BIAD (Beijing Institute of Architectural Design)
Bigen Africa Services (Pty) Ltd.
Billings Design Associates, Ltd.
bKL Architecture LLC
BluEnt
BOCA Group
Bollinger + Grohmann Ingenieure
Bosa Properties Inc.
Boston Properties, Inc.
Broadway Malyan
Brunkeberg Industriutveckling AB
Buro Ole Scheeren
Callison, LLC
Camara Consultores Arquitectura e Ingenieria
Capital Group
Cardno Haynes Whaley, Inc.
Case Foundation Company
CB Engineers
CCHRB (Chicago Committee on High-Rise Buildings)
CCHRB (Chicago Committee on High-Rise Buildings) (Inv. Billing Address)
CDC Curtain Wall Design & Consulting, Inc.
Central Scientific and Research Institute of Engineering Structures "SRC Construction"
China Academy of Building Research
China Institute of Building Standard Design & Research (CIBSDR)
City Developments Limited
Concrete Reinforcing Steel Institute (CRSI)
COOKFOX Architects
Cosentini Associates
COWI A/S
CPP Inc.
CS Associates, Inc.
CS Structural Engineering, Inc.
CTL Group
Cubic Architects
Cundall
Dam & Partners Architecten
Dar Al-Handasah (Shair & Partners)
David Engineers Ltd.
Delft University of Technology
Dennis Lau & Ng Chun Man Architects & Engineers (HK), Ltd.
Despe S.p.A.
dhk Architects (Pty) Ltd
Diar Consult
DSP Design Associates Pvt., Ltd.
Dunbar & Boardman
Earthquake Engineering Research & Test Center of Guangzhou University
EC Harris
ECSD S.r.l.
Edgett Williams Consulting Group, Inc.
Eight Partnership Ltd.
Electra Construction Ltd.
Elenberg Fraser Pty Ltd
ENAR, Envolventes Arquitectonicas
Environmental Systems Design, Inc.
Exova Warringtonfire
Farrells
Feilden Clegg Bradley Studios LLP
Fortune Shepler Saling Inc.
FXFOWLE Architects, LLP
GCAQ Ingenieros Civiles S.A.C.
GEO Global Engineering Consultants
Gilsanz Murray Steficek
M/s. Glass Wall Systems (India) Pvt. Ltd
Global Wind Technology Services (GWTS)
Glory Harvest Group Holdings Ltd
Gold Coast City Council
Goldstein Hill & West Architects, LLP
Gorproject (Urban Planning Institute of Residential and Public Buildings)
Grace Construction Products
Gravity Partnership Ltd.
Grimshaw Architects
Grupo Inmobiliario del Parque
Guoshou Yuantong Property Co. Ltd.
GVK Elevator Consulting Services, Inc.
Halvorson and Partners
Handel Architects
Heller Manus Architects
Henning Larsen Architects
Hilson Moran Partnership, Ltd.
Hines
Hong Kong Housing Authority
BSE, The Hong Kong Polytechnic University
Housing and Development Board
IECA Internacional S.A.
ingenhoven architects
Institute BelNIIS, RUE
INTEMAC, SA
Ivanhoe Cambridge
Iv-Consult b.v.
J. J. Pan and Partners, Architects and Planners
Jahn, LLC
Jaros Baum & Bolles
Jaspers-Eyers Architects
JBA Consulting Engineers, Inc.
JCE Structural Engineering Group, Inc.
JMB Realty Corporation
The John Buck Company
John Portman & Associates, Inc.
Johnson Pilton Walker Pty. Ltd.
Kalpataru Limited
KEO International Consultants
KIM-SH LLC (Complex Engineering Workshop)
Kinetica
King Saud University College of Architecture & Planning
King-Le Chang & Associates
KPFF Consulting Engineers
KPMB Architects
LBR&A Arquitectos
LCL Builds Corporation

Ledcor Construction Limited
Leigh & Orange, Ltd.
Lerch Bates, Inc.
Lerch Bates, Ltd. Europe
LMN Architects
Lobby Agency
Louie International Structural Engineers
Lyons
Mace Limited
Madeira Valentim & Alem Advogados
MADY
Magellan Development Group, LLC
Margolin Bros. Engineering & Consulting, Ltd.
Matthews Holdings Southwest Inc
James McHugh Construction Co.
Meinhardt (Thailand) Ltd.
Metropolis, LLC
Michael Blades & Associates
MKPL Architects Pte Ltd
MMM Group Limited
Moshe Tzur Architects Town Planners Ltd.
Murchie Consulting Pty Ltd
MVSA Architects
New World Development Company Limited
Nikken Sekkei, Ltd.
NPO SODIS
O'Connor Sutton Cronin
onespace unlimited inc.
Option One International, WLL
Ortiz Leon Arquitectos SLP
P&T Group
Palafox Associates
Pelli Clarke Pelli Architects
Philip Chun and Associates Pty Ltd
PLP Architecture
Porte Construtora Ltda
PositivEnergy Practice, LLC
Profica
Project and Design Research Institute "Novosibirsky Promstroyproject"
PT Anggara Architeam
PT. Prada Tata Internasional (PTI Architects)
Rafael Viñoly Architects, PC
Ramboll
Read Jones Christoffersen Ltd.
Rene Lagos Engineers
RESCON (Residential Construction Council of Ontario)
Rider Levett Bucknall North America
Riggio / Boron, Ltd.
Robin Partington & Partners
Roosevelt University – Marshall Bennett Institute of Real Estate
Safdie Architects
Sauerbruch Hutton Gesellschaft von Architekten mbH
schlaich bergermann und partner
Schock USA Inc.
Sematic SPA
Shanghai EFC Building Engineering Consultancy
Shimizu Corporation
Sino-Ocean Land
SKS Associates
SL+A International Asia Inc. Taiwan Branch
Smith and Andersen
SmithGroup
Somdoon Architects Ltd.
Southern Land Development Co., Ltd.
Sowlat Structural Engineers
Soyak Construction
Stanley D. Lindsey & Associates, Ltd.
Stauch Vorster Architects
Stephan Reinke Architects, Ltd.
Sufrin Group
Surface Design
Swinburne University
Taisei Corporation
Takenaka Corporation
Tameer Holding Investment LLC
Tandem Architects (2001) Co., Ltd.
Taylor Thomson Whitting Pty., Ltd.
TFP Farrells, Ltd.
Thermafiber, Inc.
Tianjin Jinxiao Real Estate Development Co. Ltd.
TMG Partners
TOBIN, PLLC
Transsolar
The Trump Organization
Tyréns
Umow Lai Pty Ltd
University of Maryland – Architecture Library
University of Nottingham
UralNIIProject RAACS
VDA (Van Deusen & Associates)
Vidal Arquitectos
Views On Top Pty Limited
Vipac Engineers & Scientists, Ltd.
VOA Associates, Inc.
Walsh Construction Company
Warnes Associates Co., Ltd
Waxman Govrin Geva
Web Structures Pte Ltd
Werner Sobek Group GmbH
wh-p GmbH Beratende Ingenieure
Windtech Consultants Pty., Ltd.
WOHA Architects Pte., Ltd.
Wong & Ouyang (HK), Ltd.
Wordsearch
WTM Engineers International GmbH
WZMH Architects
Y. A. Yashar Architects
Yaron Offir Engineers Ltd.
Zemun Ltd.
Ziegler Cooper Architects

Supporting Contributors are those who contribute $10,000;
Patrons: $6,000; Donors: $3,000; Contributors: $1,500; Participants: $750

T - #0512 - 071024 - C280 - 254/203/12 - PB - 9780367378196 - Gloss Lamination